P9-BYR-435

HOW THE UNIVERSE WORKS

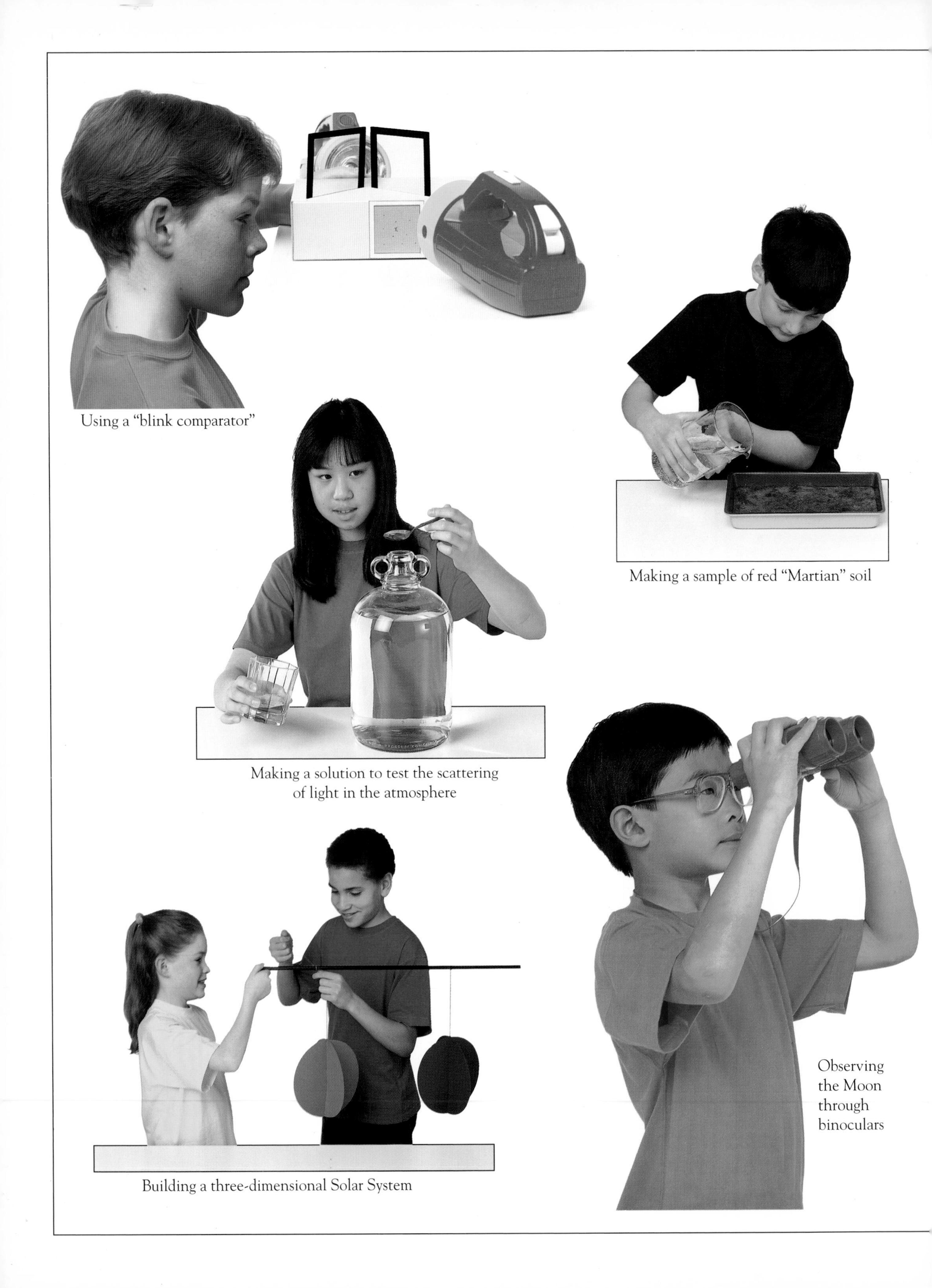

Using a "blink comparator"

Making a sample of red "Martian" soil

Making a solution to test the scattering of light in the atmosphere

Building a three-dimensional Solar System

Observing the Moon through binoculars

HOW THE UNIVERSE WORKS

Heather Couper
& Nigel Henbest

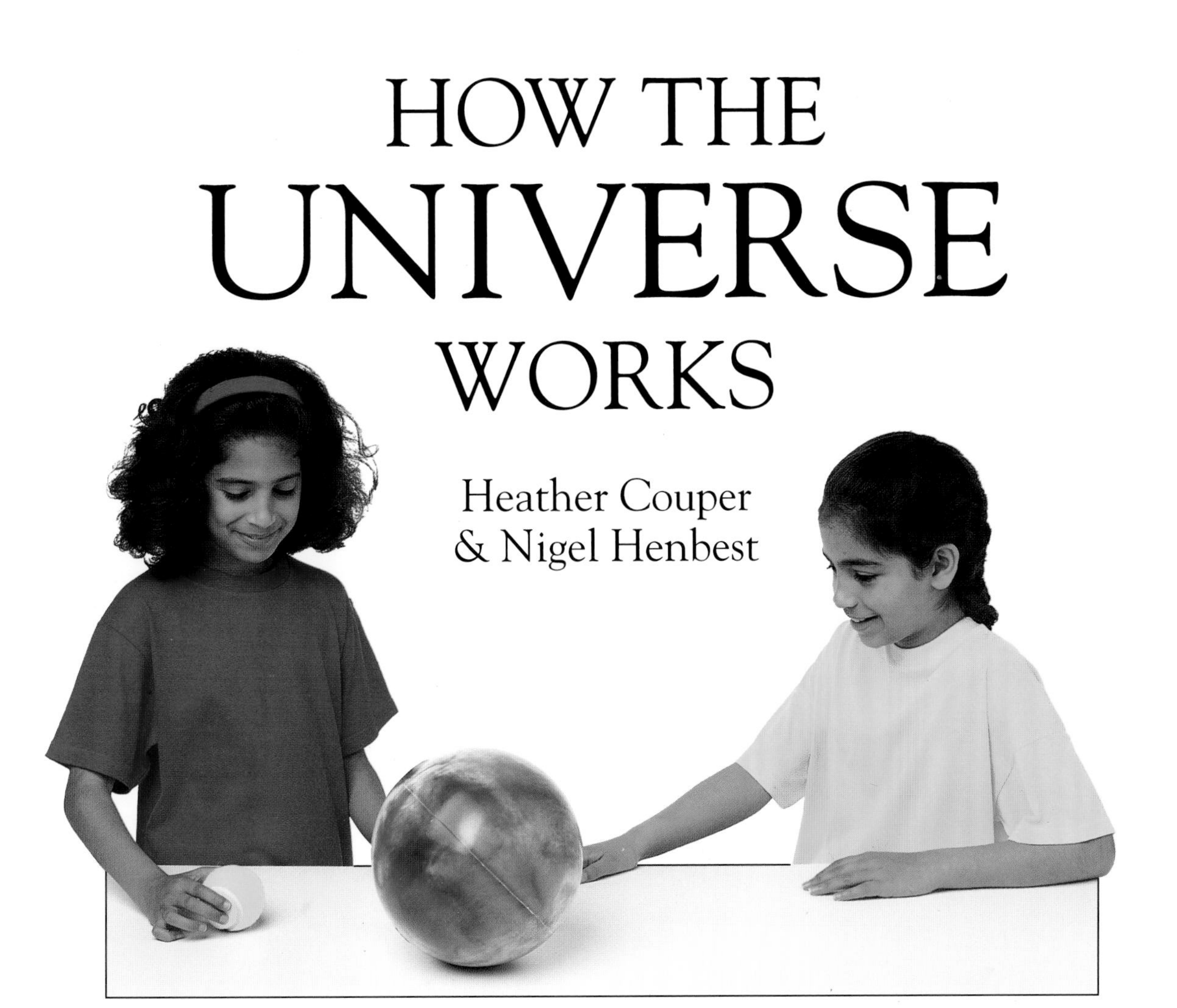

Eclipsing a "star"

Making a "black hole"

Observing the stars with a star theater

Reader's Digest

The Reader's Digest Association, Inc.
Pleasantville, New York • Montreal

A READER'S DIGEST BOOK
Designed and edited by Dorling Kindersley Limited, London

Project Editor Stephanie Jackson
Art Editors Ross George, Neville Graham, Chris Walker
Editor Nicholas Turpin
Designers Elaine Monaghan, Marianna Papachrysanthou
Production Louise Daly
Managing Editor Josephine Buchanan
Managing Art Editor Lynne Brown

The credits and acknowledgments that appear on page 160 are hereby made a part of this copyright page.

Library of Congress Cataloging in Publication Data

Couper, Heather.
How the universe works / Heather Couper & Nigel Henbest.
p. cm.
Includes index.
ISBN 0-89577-576-X
1. Astronomy—Experiments. 2. Astronomy—Laboratory manuals. 3. Scientific recreations. I. Henbest, Nigel. II. Title.
QB62. 7. C68 1994
520'.78—dc20 93- 21345

Printed in Singapore

Contents

Introduction 6
The home laboratory 8
The home observatory 10
The telescope 12

Spaceship Earth

The living planet 16
The round Earth 18
Time and place 20
The time of day 22
The time at night 24
Year after year 26
Our window into space 28
Light pollution 30
Breaking free 32
Space laboratories 34

The Moon

Earth's satellite 38
The Earth and the Moon 40
The Moon's orbit 42
Lunar eclipse 44
Moon spotting 46
Mapping the Moon 48
The Moon's surface 50
Gravity and the Moon 52
Going to the Moon 54

The Solar System

A planetary family 58
The Solar System to scale 60
Planets on the move 62
Mercury 64
Venus 66
Mars 68
Exploring Mars 70
Jupiter 72
Saturn 74
Uranus 76
Neptune 78
Pluto and Planet X 80
Planetary probes 82
Comets 84
Shooting stars 86

The Sun

Our local star 90
The Sun's energy 92
A star close up 94
The Sun's light 96
Inside the Sun 98
The solar cycle 100
Solar eclipse 102

The Stars

Starlight and star life 106
Star theater 108
The constellations 110
Dialing the stars 112
Starlight 114
How far are the stars? 116
Red giants and white dwarfs 118
Double trouble 120
Star birth 122
Star death 124
Pulsars and black holes 126

The Cosmos

Galaxies and beyond 130
The Milky Way 132
The structure of our Galaxy 134
Galaxies galore 136
Clusters of galaxies 138
Quasars 140
The expanding Universe 142
Big Bang to Big Crunch 144
Is anyone there? 146

Stars of the northern skies 148
Stars of the southern skies 150
Glossary 152
Index 156
Acknowledgments 160

INTRODUCTION

OUR ANCESTORS made practical use of astronomy in almost every aspect of their lives. They watched the Sun, Moon, stars, and planets to help them mark the passage of time. Before navigational satellites, sailors at sea used the stars in the sky as beacons to plot their course. Before modern agricultural techniques were introduced, farmers scanned the skies for seasonal clues to when they should plant and harvest their crops. Today, astronomy is still relevant to our lives. Many of the discoveries being made about the farthest reaches of our Universe help us to understand more about how our own world works.

But what does an astronomer do? An astronomer is first and foremost a physicist—someone who tries to understand the structure and behavior of matter. Instead of being confined to a laboratory, the astronomer works with the whole Universe. This means an astronomer tries to account for all the extremes of the Universe —from the enormous density of a black hole to the vacuum of space, from the searing heat of

the Big Bang to the cold of the planet Pluto. But astronomy is more than fundamental physics. In exploring the Universe, you will come up against some "big questions." How did the the Universe begin? How big is space? Will the Sun ever die? Is there life elsewhere in the Universe? The answers to these questions lie not only in astronomy, but also in mathematics, biology, chemistry, earth science, information technology, and physics. In short, astronomy is a gateway to all the sciences.

In *How the Universe Works* you will learn about the Universe through simple experiments, using items easily found in your home. Find out how to measure the heights of mountains on cloud-covered Venus, estimate the temperature of a star, make your own telescope, and learn how to combat light pollution in your neighborhood. Even very young children can join in the fun—we have devised experiments for all ages and indicated the need for adult supervision where appropriate. Although you may not go on to become a professional astronomer, we hope that this book will give you a better idea of what makes the Universe "tick." Most of all, we hope that you get to know the planets and stars, and all the other objects in the Universe.

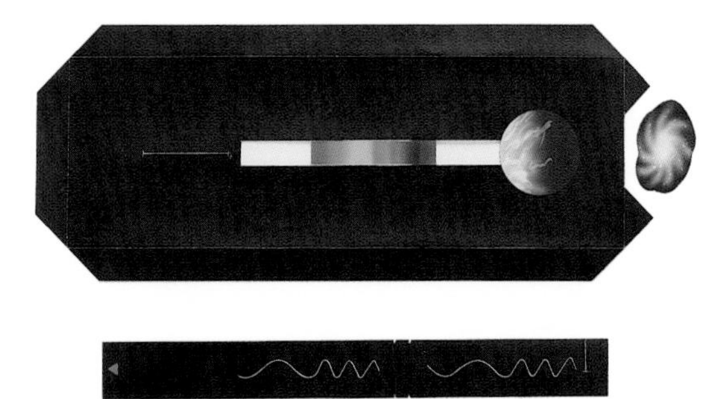

The home laboratory

PRACTICALLY ALL the experiments shown in this book can be easily performed with simple materials, tools, and utensils. Most of the things you will need can be found in and around the home, generally in the kitchen, or they can be acquired easily and cheaply from somewhere else. On these pages we show you some of the items that you will find useful in putting together your own astronomy laboratory. If you don't have what is shown here, try using something similar—it may work just as well.

Mixers and measurers

Accurate measurements are needed in many everyday activities as well as in scientific experiments. Most scientists now use the International System (SI), or metric system, for measures, with meters for length, kilograms for mass, and seconds for time. We have shown both common and metric measurements (sometimes rounded off to be easier to use), but be sure to use only one measurement system throughout each experiment. Here are some items you may find useful for the mixing and measuring that you will do.

A watch or clock with a second hand

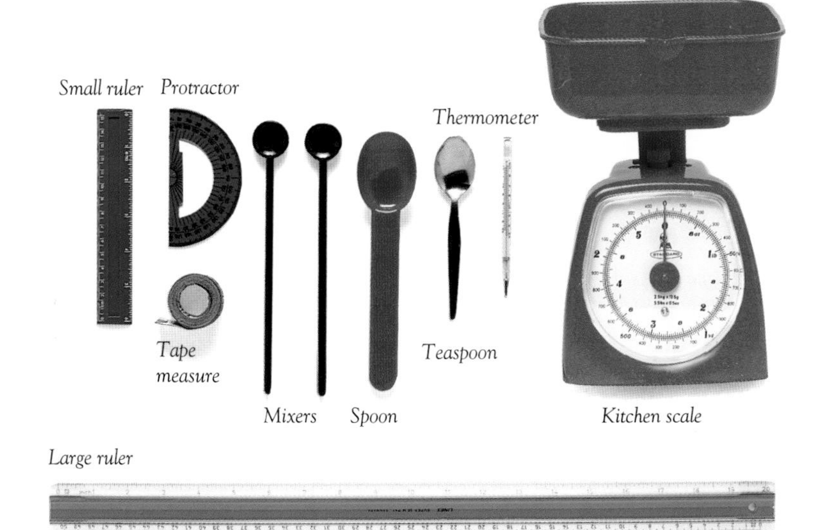

Electrical items

To make a simple circuit, tape the end of an electrical wire to the top of a battery. Connect the other end of the wire to a light bulb. Connect another length of wire to the bulb, and tape that to the bottom of the battery. You will now be able to switch on the light bulb.

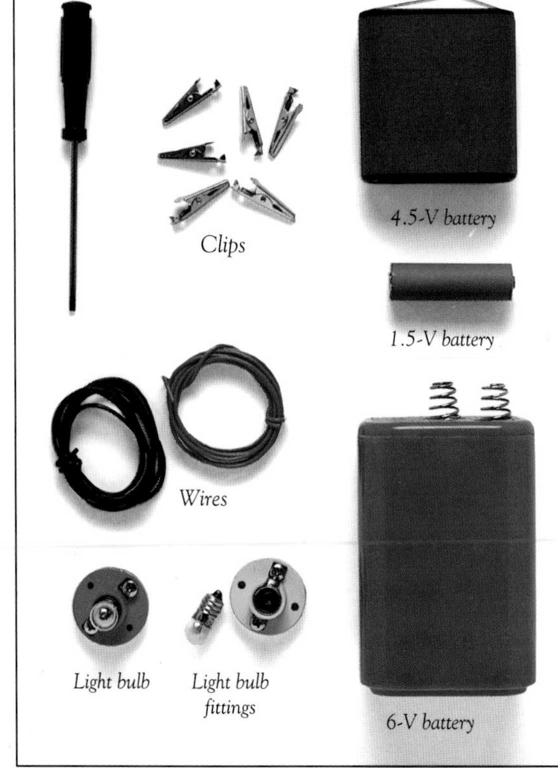

Equipment

You will need scissors, pens, pencils, and paints for many experiments. Ask an adult to help with any project that will involve the use of sharp tools or instruments.

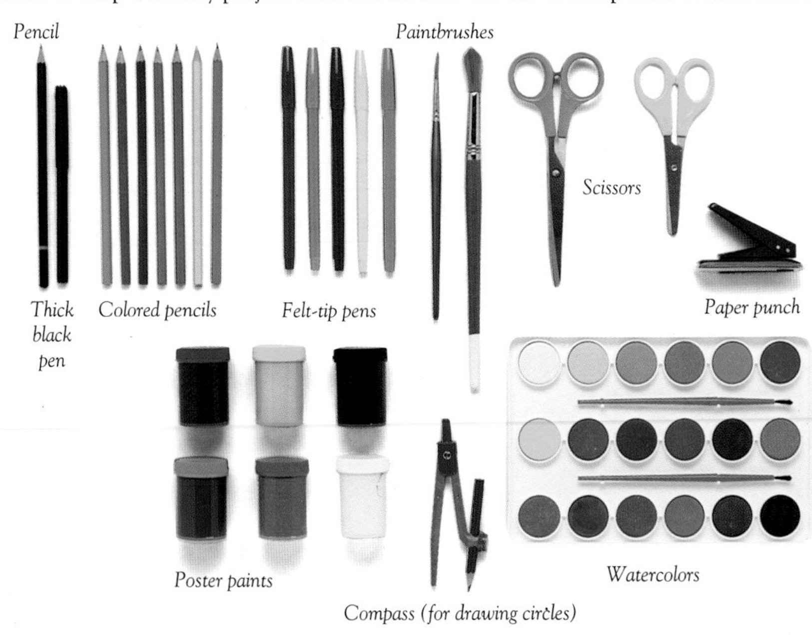

Materials

For some of the experiments and projects in this book, you will need to buy a few basic materials, such as paper, modeling clay, or poster board. Foamcore, which we use to construct many of the experiments, is a piece of of foam that is sandwiched between two pieces of poster board. It is available in art-supply stores.

Useful items

All kinds of items from around the house—even old dishwashing gloves and corks—may be used for some of the experiments. Be sure to ask permission first before taking something away.

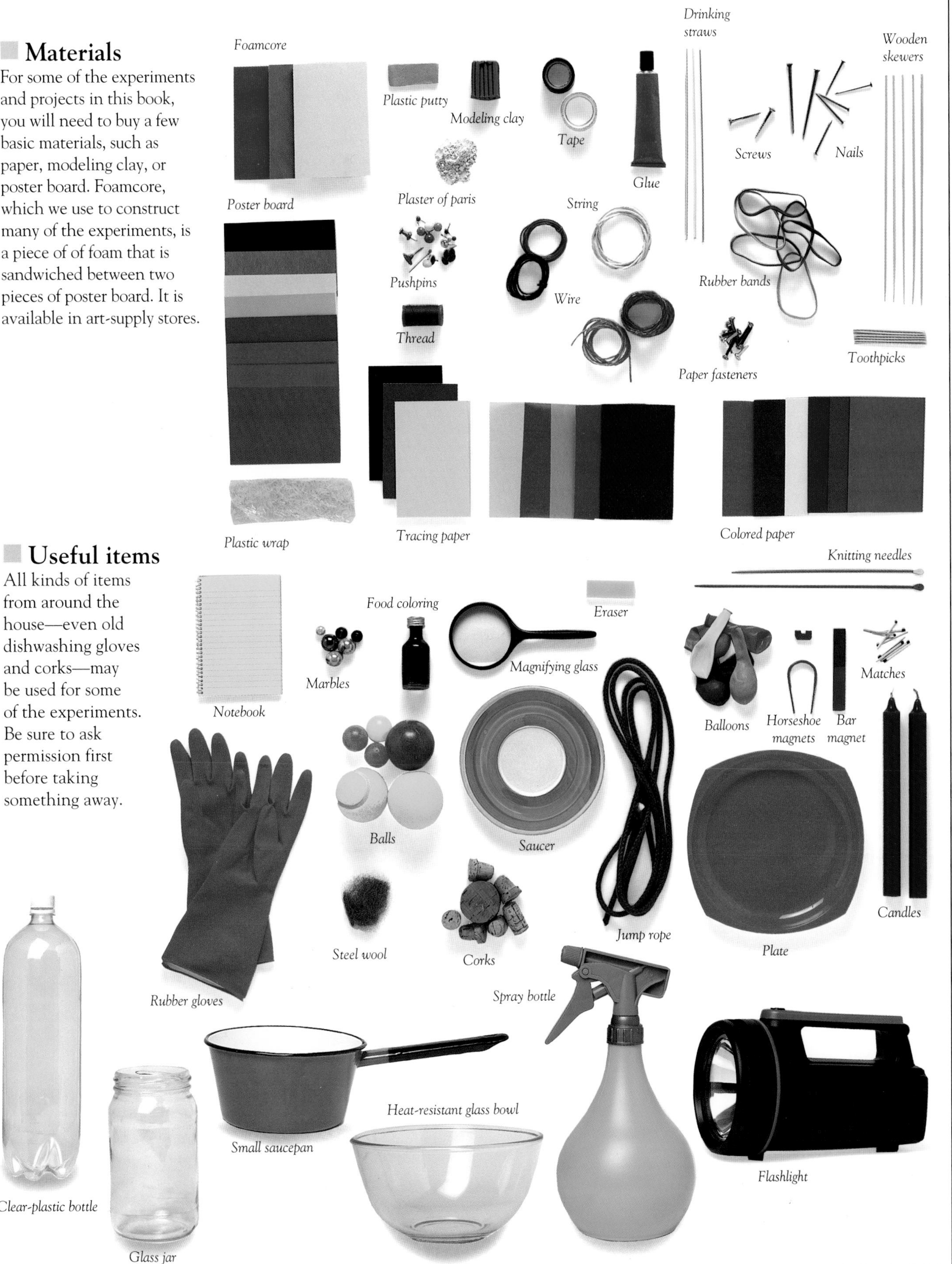

The home observatory

MOST SCIENCE CAN be studied right in your home laboratory (pp.8–9), but astronomy is different. To study what is happening in the Universe you need to go out and look at the sky. On any clear day you can see the Sun. On any clear night you can see some of the stars—and the Moon, if it is up. You can view the heavens through a window, but by setting up a home observatory you will have a place where you can comfortably observe and record information about celestial objects. The word "observatory" may conjure up the image of a giant telescope in a huge dome, but an observatory is simply any place from which you can observe. For your home observatory, all you need is an outdoor spot that is safe to use and that gives you a good view of the sky. Your observatory will become more than just a place from which to see; it will also be a place for you to analyze, evaluate, monitor, and record.

Dark adaptation

When you go outside at night to observe the sky, give your eyes some time to adjust to the darkness. It takes nearly half an hour for your eyes to get used to the dark, enabling you to see the faintest stars. A bright light—like a flashlight used to illuminate your notes and star maps—can ruin your "dark adaptation" instantly. To remain adapted to the dark, tape red cellophane over your flashlight beforehand. Red light is far less damaging to your dark adaptation than white light.

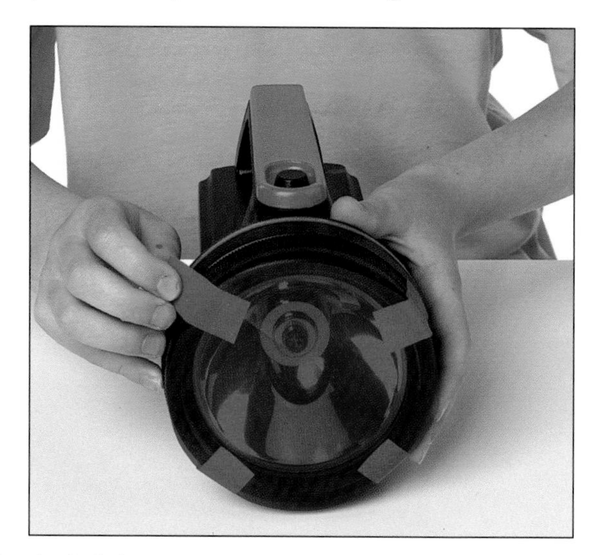

Cover a flashlight with red cellophane to help you see in the dark

Setting up your home observatory

Although you may do much of your observing at night, it is best to find your observatory site during the day, when you can see what you are doing. Look for a place that has a wide view of the sky, especially toward the west and south if you are in the Northern Hemisphere (or the north if you are in the Southern Hemisphere), and a flat space where you can observe. The essential items for your observatory are simply a chair to sit on, a table, and a reclining chair. Draw the compass points (N, S, E, W) on a piece of poster board, and pin this to the table in a clear-plastic bag, so that "S" points toward the position of the midday Sun (or "N" in the Southern Hemisphere). Put a chair at the north end of the table (in the Southern Hemisphere, the south end). Your home observatory is now ready. Always be sure an adult knows that you are using it.

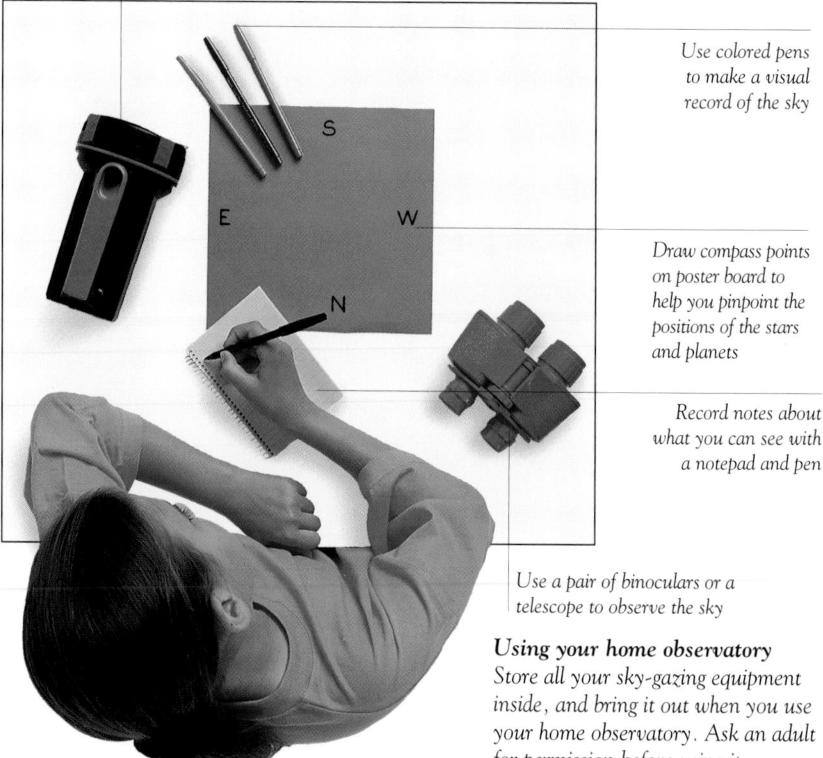

Use colored pens to make a visual record of the sky

Draw compass points on poster board to help you pinpoint the positions of the stars and planets

Record notes about what you can see with a notepad and pen

Use a pair of binoculars or a telescope to observe the sky

Using your home observatory
Store all your sky-gazing equipment inside, and bring it out when you use your home observatory. Ask an adult for permission before using it.

Rounding off numbers

Many of the measurements in this book, such as the distances between stars, have been rounded off. In astronomy this is standard practice, though it is not in mathematics.

The sky at night

To make the most of your time spent observing the sky at night, first go outside to your home observatory, switch on a flashlight, and look at the compass points to become familiar with directions. Use a star-dial (pp.112–113) and a newspaper or almanac to identify the brightest stars and any planets that may be visible. Over the next 10 to 20 minutes, you will be able to see fainter and fainter stars. Now you can move to a more comfortable position in a reclining chair. Keep any items that you will need nearby. Try to identify the fainter stars, star clusters, nebulae, and galaxies in the star charts of this book (pp.148–151). Also look out for unexpected sights such as shooting stars and the auroras (known as the aurora borealis in the Northern Hemisphere and the aurora australis in the Southern Hemisphere).

Observe the stars close up with binoculars or a telescope—be sure to hold them firmly

Draw the stars you can see

Time meteor showers with a watch with a timer

Wrap up in a sleeping bag to keep warm

Use star charts for handy reference

Sit in a comfortable reclining chair

Bring a thermos with a hot drink to help keep you warm

Take pictures with a 35-mm camera

Use a flashlight to check your notes

Monitor direction with a compass

Photographing the sky

You can photograph the sky during the day just as you would take an ordinary photograph. To take pictures at night you need a manually operated 35-mm camera. Mount it on a tripod to keep it steady, or hold the camera against a firm surface. Experiment with exposure times—the brightest planets and stars will show up on an exposure of 10 to 30 seconds, and longer exposures will show fainter stars. Keep notes on how you take every shot. Take occasional daylight photographs on the roll of film you use for night photography, especially the first picture, and label the roll "astronomy photos—print everything" when you take it in for processing. Otherwise, the developers may think you have a blank roll of film.

The Moon in daytime
This photograph was taken in the daytime. A daytime shot is much easier to take than a nighttime photograph because there is less glare, allowing more detail on the Moon to be seen.

The stars at night
When photographing the night sky, bracket each shot: take one shot you think is right, one at half the shutter speed, and one at twice the speed.

The telescope

ASTRONOMERS USE TELESCOPES to investigate the Universe. Telescopes magnify distant scenes that the eye cannot make out. For example, a planet looks merely like a bright star to the naked eye, but through a telescope details such as the spots on Jupiter, the markings on the deserts of Mars, and the rings of Saturn can be seen. Even more important than magnifying, a telescope can capture more light than the eye on its own, so you can see fainter and more distant objects than are visible with the naked eye. You can either buy a telescope or make one. Alternatively, you can use a pair of binoculars.

EXPERIMENT

Making a telescope

If you do not already have a telescope of your own, you can make a simple one. If it is assembled with care, your telescope will actually outperform some of the telescopes that you can buy in stores. It will show you the craters on the Moon, the bulging equator of Jupiter, and possibly the rings of Saturn. It will also reveal double stars, star clusters, and nebulae ("star factories"). For the best results with any telescope, support it firmly. You can do this by resting your elbows on the tabletop of your home observatory (pp.10–11) and holding the telescope steady, or you can mount your telescope on a stand, such as a camera tripod, with a lump of plastic putty to fix it. When you are using your telescope, bear in mind that the images you see will be upside down. But this does not matter, because there is no "up" or "down" in space.

YOU WILL NEED

- *For Lens 1 (the eyepiece) use a "loupe"—a special lens used to examine jewels and 35-mm photographic slides. Any good camera store will stock one.*

- *For Lens 2 (the objective), ask an optician for a convex 70-mm lens with a power of 2 diopters. If possible, ask the optician to coat the lens with anti-glare solution.*
- *thick black paper*
- *thick black tape*
- *compass with a white pencil*
- *stick-on stars for decoration*
- *large ruler*
- *scissors*
- *glue*

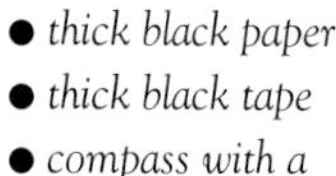

The lenses of your telescope

If you cannot obtain lenses exactly like the ones specified above, you can make some substitutions. For Lens 1 (the eyepiece) you need a small lens with a short focal length (about 1¼ in [3 cm]) so that the image can be brought into focus in the short space between the eyepiece and the back of your eyeball. For Lens 2 (the objective) you need a convex (magnifying) lens about 2¾ in (70 mm) wide with a long focal length—about 20 in (50 cm)—to send the image down the telescope's tube. To check the focal length, focus the Sun through the lens onto paper. When the image is sharp, measure the distance between the lens and the paper—this is the focal length. Tubes A and B should each be three-fourths of the focal length, so adjust their length to suit your lens.

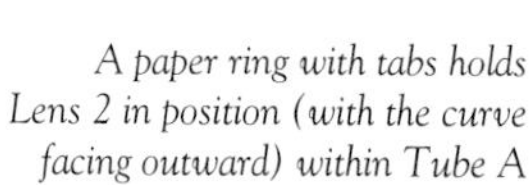

A paper ring with tabs holds Lens 2 in position (with the curve facing outward) within Tube A

Lens 2 is the objective lens—make sure it is clean before you assemble the telescope

A ring cut from black paper is glued to Lens 2

Tube A should be three-fourths the focal length of Lens 2; the inside diameter equals the diameter of Lens 2

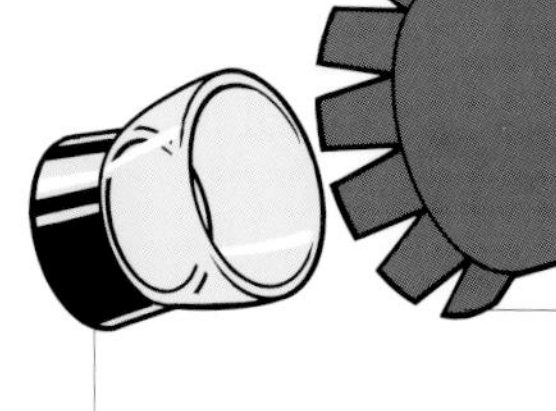

A loupe is ideal for Lens 1 (the eyepiece) because it has a solid frame to which Tube B can be firmly attached—if you use another type of lens, you need to adjust Tube B to fit around it

Cut tabs into the end of Tube B; bend and firmly tape these around Lens 1

The tubes are made of rolled black paper and are the same length, but Tube B has a slightly smaller diameter so that it fits snugly into Tube A

Buying a telescope

If you want to buy a telescope, you will find that several different kinds are available. The type of telescope that has a lens at the front (a "refractor") is affordable in smaller sizes, but it becomes more costly as it gets bigger. If you want a more powerful telescope that is not too expensive, go for a "reflector": this has a curved mirror at the bottom to collect and focus light. The best are the "catadioptric" telescopes, which have both a weak lens at the front and a curved mirror. They are very powerful, but not too large or cumbersome.

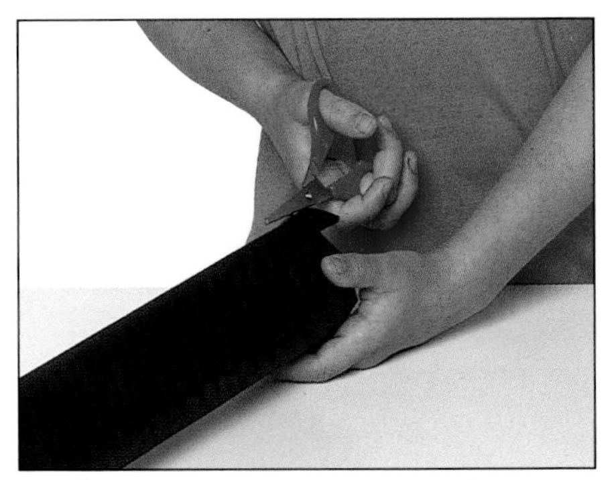

1 CUT OUT NARROW WEDGES about 1 in (25 mm) long all around the end of Tube B (see opposite), so that you end up with tabs that can be bent inward.

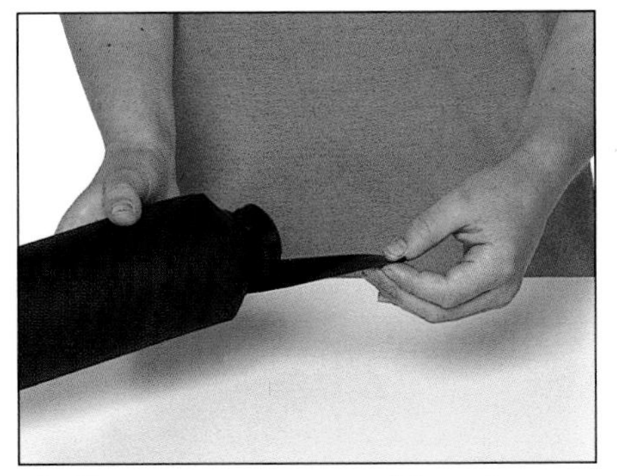

2 BEND THE TABS around Lens 1, keeping as even a circle as you can around the width of the tube. Tape them down to the lens so that it is firmly fixed.

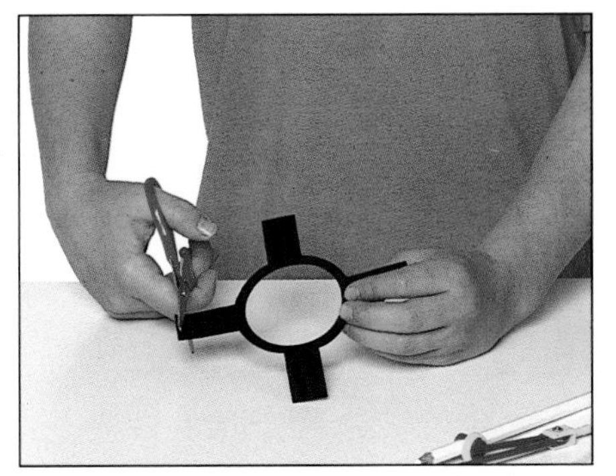

3 CUT OUT A BLACK RING with tabs. The outer diameter of the ring should match the diameter of Lens 2. Make the tabs 1 in (25 mm) long.

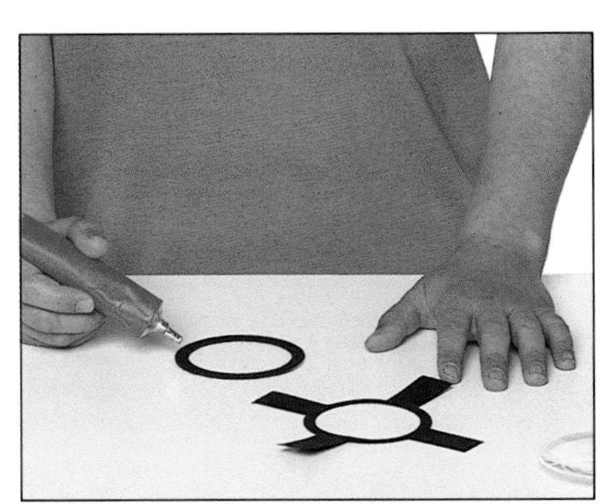

4 TRACE AND CUT out a plain black ring the same size as in Step 3. Glue Lens 2 to it, and glue it with the lens into the middle of the ring with the tabs.

5 FOLD THE TABS forward and with the tabs pointing toward you, slide Lens 2 into Tube A. Glue the tabs so that the ends are level with the tube's end.

Add stick-on stars for decoration

6 CLEAN ANY FINGER marks off the lenses, slide Tube B into Tube A, and your telescope is complete. Rest your elbows on a firm surface, or rest the telescope on something solid, like a windowsill, to keep it steady. To start, focus the telescope on the Moon (the best time to view is at half Moon)—slide Tube B inside Tube A until the view is sharp.

Binoculars

Binoculars are very useful for getting started in astronomy. A pair of binoculars is like having two telescopes side by side, so you can use both eyes at once. The view through binoculars is the right way up, so you can use them for other hobbies besides astronomy. Binoculars are labeled with two numbers (7 x 50, for example). The first is the magnification, the second the diameter of the objective lenses in millimeters.

Why telescopes give a flipped image

Lens 2 bends light rays in. The point where the rays meet is the focal point, and the distance from there to Lens 2 is the focal length. Beyond the focal point, the rays spread out again, producing an upside-down image. Lens 1 acts like a magnifying glass, enlarging the image without flipping it over.

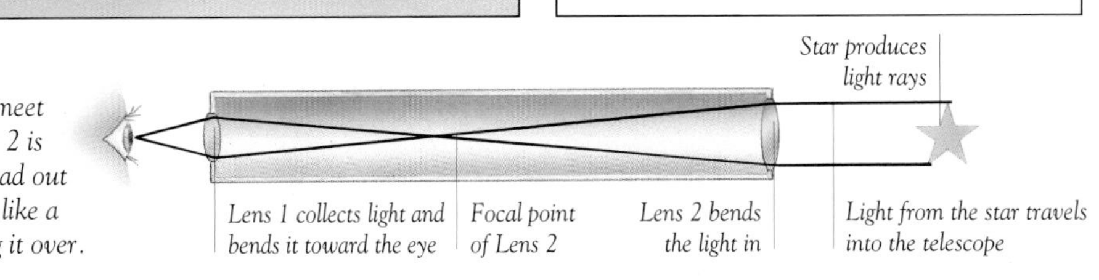

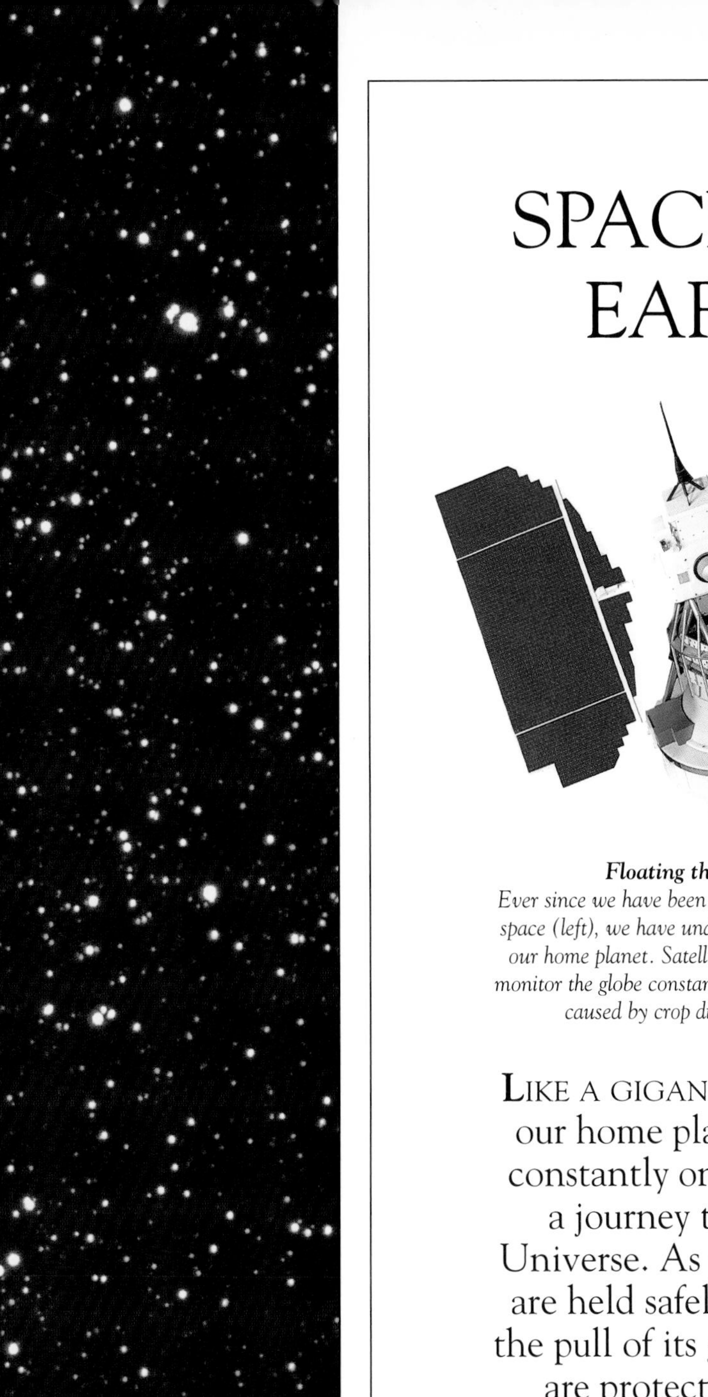

SPACESHIP EARTH

Floating through space
Ever since we have been able to look at Earth from space (left), we have understood much more about our home planet. Satellites like Landsat *(above) monitor the globe constantly and can detect changes caused by crop disease or pollution.*

LIKE A GIGANTIC SPACESHIP, our home planet, Earth, is constantly on the move on a journey through the Universe. As passengers, we are held safely on board by the pull of its gravity, and we are protected from the harshness of space by Earth's life-support systems. But now we are starting to use our planet as a space platform—a springboard from which we can explore the worlds beyond our own.

THE LIVING PLANET

WE ARE ALL SPACE TRAVELERS. You don't need a space shuttle or a starship to speed through the Universe. We live aboard a huge spaceship whizzing through the Cosmos 30 times faster than a rifle bullet. Our spaceship is called Earth. The Earth is a ball of rock 7,972 miles (12,756 km) wide. The force that holds us to the planet's surface, gravity, also holds on to a layer of air we can breathe. This atmosphere is our life-support system and our protection from the hazards of space.

Our atmosphere
This photograph of the Earth was taken by Apollo 11. It shows that at sunset the blue part of sunlight is absorbed by the atmosphere, leaving a pink light. Our atmosphere is the protective shield between the surface of the planet and the extremes of space beyond.

The Earth is round, and gravity always pulls us toward its center. This means that people on the other side of Earth will not fall off, even though they seem to be upside down to us. People on different parts of the globe look out into space in different directions, and they can see different stars. However, you don't have to go abroad to see the stars change.

Like clockwork

The Earth spins around once every 24 hours, so our view of the Universe gradually shifts at night. At the same time, our spaceship is hurtling around the Sun at almost 68,350 mph (110,000 km/h). As we go around, we look out in different directions in space, in the same way you see the different booths at a fair when you ride a carousel. The Earth's orbit lets us see different constellations at different times of year. For example, Orion, the mighty hunter, dominates the sky in January and February, while his place is taken in July by Scorpius, the Scorpion—which, in Greek myth, killed Orion by stinging his heel.

Turbulent seas
These waves are breaking off Hawaii in the Pacific Ocean. The Earth is unique in having oceans of water. Waves are whipped up by winds—air moving in the atmosphere.

Since prehistoric times, people have used the positions of the stars in the sky to tell them the time of night and the date of the year, and to navigate around the Earth. The early Polynesians traveled across thousands of miles of the empty Pacific Ocean, relying on their knowledge of the stars.

The stars are around us all the time, but we cannot see them during the daytime because the Sun is so brilliant that it lights up the atmosphere, producing a bright blue sky. Although the atmosphere can be a nuisance to stargazers—especially when it is cloudy—it is an essential part of the life-support system of our spaceship Earth.

Fossil remains
Life has been present on the Earth for billions of years, as we can see from the study of fossils such as this Trilobites sphaerexochus. *So far, there is no evidence that life exists anywhere else in the Universe except Earth.*

Watery world

Without the atmosphere, we would be exposed to all the extremes of space. The side of the Earth facing the Sun would be baked; the other side frozen. There would be no protection from meteors—small pieces of rock hurtling in from space —or from dangerous high-energy radiation from the Sun. Living beings would have nothing to breathe. There would be no winds or weather systems and, most important, no water. If there were no atmosphere to press down on an ocean or a lake, the liquid water would just boil away into the vacuum. For similar reasons, astronauts in space must wear air-filled pressure suits to prevent their blood from boiling. Two-thirds of the Earth is covered in liquid water, which makes it unique among the worlds in our Solar System. Astronauts orbiting the Earth high up, and seeing its extensive oceans, call it the "Blue Planet."

Lush life
Why is it that the Earth can support life and the other planets cannot? The huge variety of life on Earth is incredible—even today, new types of plant and animal life are still being discovered. However, many species are dying out, and it is up to us to protect them.

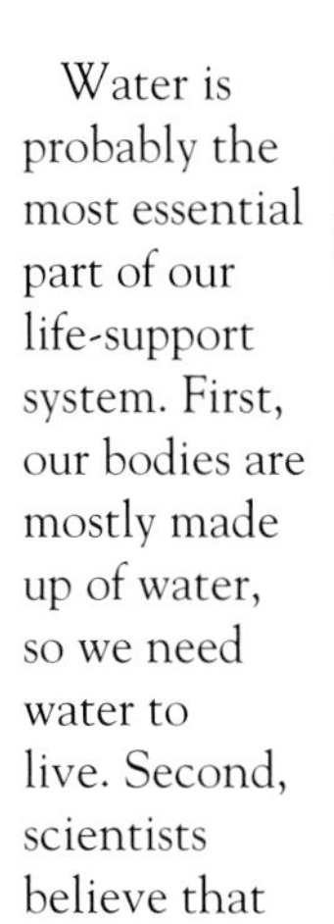

Spitting fire
Volcanoes are a constant reminder of the Earth's regenerative power.

Water is probably the most essential part of our life-support system. First, our bodies are mostly made up of water, so we need water to live. Second, scientists believe that water was crucial in helping to get life started on Earth. The complex chemical reactions that gave rise to life, and continued to support it, needed water to sustain them.

Life on Earth

Spaceship Earth carries an enormous crew on its journey through the Cosmos. From people to plants, from amoebas to aardvarks, our planet is teeming with life. The different forms of life help each other to survive. Plants give off oxygen gas, which animals then breathe. Animals produce waste that fertilizes the ground and helps plants grow. Some scientists regard all living things on Earth as making up a single, vast living entity called Gaia ("Mother Earth" in Greek mythology).

How life began is a topic still hotly debated by scientists. Some believe there is evidence that the building blocks of life were brought to Earth by comets. Others think that lightning flashes in the early atmosphere formed the raw materials. What is certain is that life started very quickly after the Earth was born. The fossils of the earliest life-forms are 3.8 billion years old, which means that they came into being only 800 million years after our planet's formation.

Folding plates
Machhapuchhare, a mountain in the Himalayan mountain range, was formed by plate tectonics.

Life is another feature that makes the Earth unique. As far as we know, no other planet in the Solar System has living creatures. It seems that the Earth had just the right combination of conditions for life to form. Most important, it is at exactly the right distance from the Sun to provide the right environment for life. Not too hot and not too cold, the Earth has not been baked like Venus or frozen like Mars. Just a bit closer to or farther from the Sun, and it would have been a different story.

As well as teeming with life, the Earth is, in a sense, a living planet. Like its neighbors, it is a relatively small, rocky world. But the Earth is slowly and constantly renewing and replenishing itself. Its surface, the crust, is made of huge plates, which slowly float around the globe on a "sea" of hot, molten rock that lies beneath. As the plates move, the continents gradually drift around the globe—something unique to our planet. When the plates collide, molten rock wells up, creating new crustal material —along with volcanoes and earthquakes. The slow movement of the plates (plate tectonics) crumples up the crust to create huge mountain ranges.

Interior space

Inside, too, the Earth is active. It has a solid inner core made of iron at a temperature of 7,200° F (4,000° C), which is surrounded by a layer of liquid iron. Currents in this fast-spinning liquid generate the Earth's magnetic field —the force that is responsible for making compass needles point in a particular direction. This magnetic field extends into space, creating a protective bubble around our planet and channeling electrically charged particles from the Sun to the polar regions. When these hit the upper atmosphere, they create a beautiful red and green glow in the sky—an aurora.

Night and day
The Earth rotates at a steady rate. When it is daytime somewhere on the Earth, it is night somewhere else. You can show this with your own model of the Earth, by rotating it on its axis (pp.20–21).

Our planet is unique in the Solar System, but could there be other "earths" around other stars? People are looking, and one day we may travel to these other worlds. Recently, we have started to break free of our spaceship. Satellites and space stations orbit the Earth. People have walked on the Moon. Uncrewed spacecraft have explored the planets. Space engineers are now working on plans for interstellar spacecraft to travel to the nearest stars.

Magnetic magic
The auroras that can be seen near the poles are often very dramatic. This photograph shows the aurora australis, photographed from Australia. The auroras occur when atomic particles from the Sun follow the Earth's magnetism and crash into the atmosphere.

Spacewalker
Astronaut Bruce McCandless is shown here "spacewalking" a few yards from the space shuttle. Although he is in space, his orbit around the Earth is just a first step into the vast Universe.

The round Earth

MOST PEOPLE TODAY know that the Earth is round because they have seen a globe or photographs of the Earth that were taken from space. Because of the Earth's round shape, you can pinpoint your location north or south of the Equator by looking at the stars (see below). Over the centuries this has been very helpful to navigators trying to find their way on the open sea. During the 20th century scientists have been able to discover what is inside the round Earth. They have probed the planet's interior and learned that the Earth is much more than just a ball of rock.

EXPERIMENT

Find your latitude

Polaris, or the Pole Star, is a star almost directly over the North Pole. People at the North Pole see Polaris directly overhead. The farther away from the North Pole you travel, the lower Polaris seems to sink in the sky. In the North you can use Polaris to find your latitude (the distance from the Equator) because the altitude (p.152) of Polaris is equal to the latitude of the place from which it is observed. Polaris is not visible in the Southern Hemisphere, but the Southern Cross (Crux) can be used to find latitude at certain times. Wherever you live, you can make this apparatus to find your latitude.

YOU WILL NEED

- *poster board* ● *scissors* ● *compass*
- *pencil* ● *5¼ x 1¼ x ¾ in (135 x 30 x 20 mm) wood* ● *glue* ● *paper* ● *string*
- *pushpins* ● *protractor* ● *pens*
- *ruler* ● *plastic putty* ● *tape*

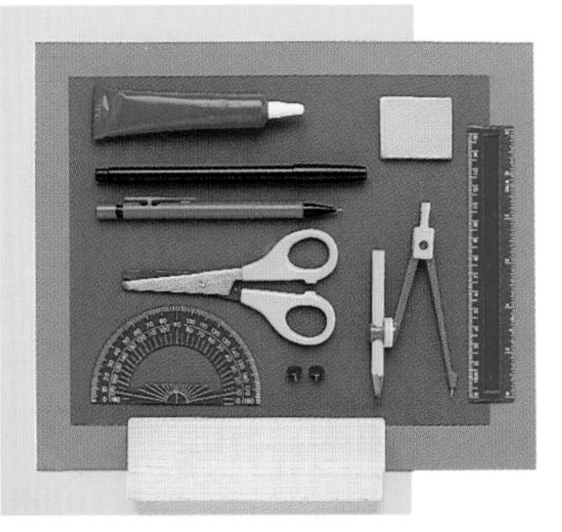

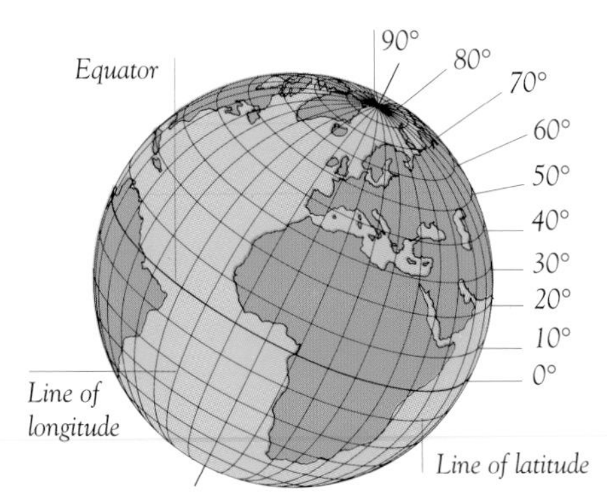

The lines of latitude and longitude
You can get an idea of your latitude from a globe. The Equator is at 0° latitude, and the poles are at 90°. Latitude lines run from east to west. The lines running from north to south are longitude.

1 DRAW A QUARTER CIRCLE, with a radius of 4½ in (114 mm) with borders on the straight sides: one ¾ in (20 mm) deep; the other 2¼ in (60 mm) deep—fold this border in half.

2 FOLD BACK the outer 1½ in (40 mm) of the border. Measure the poster board's curve with string—it should be 7¼ in (185 mm). Mark every 1/16 in (2 mm), with a larger mark every ¾ in (20 mm).

EXPERIMENT

The Earth as a magnet

The Earth has a spinning, fluid iron core that generates a strong magnetic field. You can detect this with a compass or anything that has been magnetized, such as a nail.

YOU WILL NEED

- *2-in (5-cm) nail* ● *horseshoe magnet*
- *ruler* ● *plastic putty* ● *compass*
- *1½-ft (45-cm) thread*

Earth

Earth's magnetic field

Making a magnet
Stroke a nail several times in one direction with one pole of a horseshoe magnet. Continue for 2 minutes to magnetize the nail. Balance the nail on a ruler (held on its side with plastic putty) to find the nail's center of gravity. Tie the thread at this point so that it balances, and suspend the nail so that it can swing freely. When the thread stops moving, note where the nail is pointing. Then gently tap the nail—in which direction does it settle? What force is acting on the nail?

3 LABEL THE MARKS from the bottom of the curve as degrees, with 0° at the bottom and 90° at the top. Then glue the wood to the poster board, inside the fold of the larger flap.

4 CUT SOME poster board into a 6 x ¾ in (150 x 20 mm) rectangle with a point at one end. Draw a line along the center. At 4½ in (114 mm) from the point, stick a pushpin through to the corner of the quarter circle and into the wood. This makes a pivot.

5 WRAP SOME PAPER around a pencil, and fasten it with tape to make a tube slightly shorter than the pointer. Glue the tube along the line on the pointer. Place the finished apparatus on the edge of the table in your home observatory (pp.10–11) so that the pivot is at one corner of the table and the entire apparatus is level.

90 80 70 60 50 40 20 10 0

Pointer

Sight through the tube

Using Polaris to find latitude

Point the apparatus due north. Identify Polaris (see p.24), and line up the apparatus with the star. Look through the tube, and pivot it until you see Polaris in the center. Hold the pointer in position, and check the curved scale to see where the tip of the pointer falls. This is the altitude of Polaris, and it is equal to your latitude.

Using the Southern Cross to find latitude

The Southern Cross can be used to find latitude during certain hours on certain days, when it appears to be either upright or upside down. For midevening, this means April (upright) or October (upside down). Measure the altitude of Acrux, the star at the base of the constellation. If the Cross is upright, subtract 27° to find your latitude; if it is upside down, add 27°.

Sighting the stars

For centuries, people have used more refined versions of the apparatus above to aid navigation. They come in many forms. This one—a quadrant—has a small handle at the bottom to make it easier to hold.

EXPERIMENT

Proving that the Earth is round

If you ever visit the coast, watch the ships as they travel toward the horizon. If the Earth were flat, a ship would grow smaller as it sailed away and never disappear. In fact, the lower parts of a ship will gradually vanish—hidden by the Earth's curvature—until just the top of the ship stands above the horizon. You can test this on flat and curved surfaces, using a flag to represent a ship.

YOU WILL NEED

- *paper* • *scissors* • *tape*
- *toothpicks* • *ball*

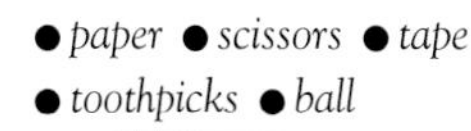

1 MAKE TWO FLAGS out of the paper, tape, and toothpicks. Move one flag farther and farther away from you along a flat table. What do you see? What happens to the flag?

2 NOW PLACE the other flag in front of you on the top of the ball, and slowly move the flag away from you. Can you see the entire flag as it moves backward over the ball?

Time and place

TO FIND EXACTLY WHERE YOU ARE on the world, you need to know two things: how far north or south of the Equator you are and how far to the east or west you are. The first part—your latitude—can be measured using the stars (pp.18–19). Finding your longitude (the distance east or west of a given point) is more difficult. Longitude is measured from the observatory at Greenwich in London, in England. The fact that the Earth spins once every 24 hours helps us determine longitude. If the Sun is overhead at your house 6 hours after it is over Greenwich, you must be one-fourth the way around the world from Greenwich, since 6 is one-fourth of 24. In this way you can use time to find your place on Earth.

EXPERIMENT

Our spinning planet

This experiment shows that when it is daytime in one place on Earth, it is night somewhere else. As the Earth spins on its axis, certain parts face the Sun, while other parts turn away from the Sun.

YOU WILL NEED
- *rubber ball* • *wooden skewer* • *drinking straw*
- *scissors* • *poster board* • *compass*
- *ruler* • *pencil* • *glue*
- *pushpins*
- *lamp*
- *paintbrush*
- *paints*

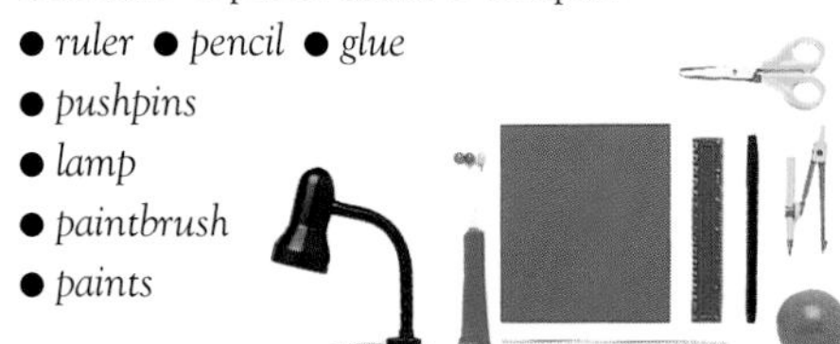

1 PIERCE THE MIDDLE of a soft rubber ball with a skewer. The ball represents the Earth, and the skewer represents the axis of the Earth.

2 USE YOUR SCISSORS to cut two short pieces of drinking straw. Slip these over each end of the axis, so that they cover the skewer entirely.

3 HOLD together two poster-board rectangles (about 3 in [75 mm] wider than the ball), and draw a circle 1 in (25 mm) wider than the ball. Cut it out.

4 BEND BACK the bottom 1½ in (4 cm) of each rectangle for a base. Put the "Earth," tilted on its axis, between the rectangles. Glue the rectangles together to hold the axis. The globe now turns freely.

5 IF YOU LIKE, paint the ball to look like the Earth. Write "midnight" on one side of the poster board and "noon" on the other. Stick a colored pushpin in the globe to mark where you live.

6 SHINE a lamp (the "Sun") to face "noon." Turn the globe counter-clockwise (from the top). When the pin is at "noon," the "Sun" shines directly on it. Move the globe so the pin is in the unlit region—the "Sun" has set. As "Earth" moves, this pattern continues.

Time differences

Different places around the world face the Sun at different times, so they must set their clocks to different times. Places with the same time form a "time zone"—a narrow strip that runs from north to south. There are 24 time zones around the world, beginning at Greenwich (see below). If you move from one zone to another, you must change your watch by 1 hour, though there are a few places that have only ½-hour or ¾-hour time differences from their neighboring time zones. If you cross the International Date Line, the date changes. There are some easy ways to find the difference in time between two places: by looking at a map that shows time zones (below), by phoning a friend who lives far away and asking him or her the time, or by tuning in to an international radio station that broadcasts the time at Greenwich. Once you have found the difference in time between one place and another, it is a simple process to find the difference in longitude. Simply multiply the difference in hours by 15 (1 hour is equal to 15° of longitude). If you are 8 hours from Greenwich, for example, the difference in your longitude is 120° If you are on Daylight Saving Time (p.22), remember to subtract 1 hour first, before you multiply.

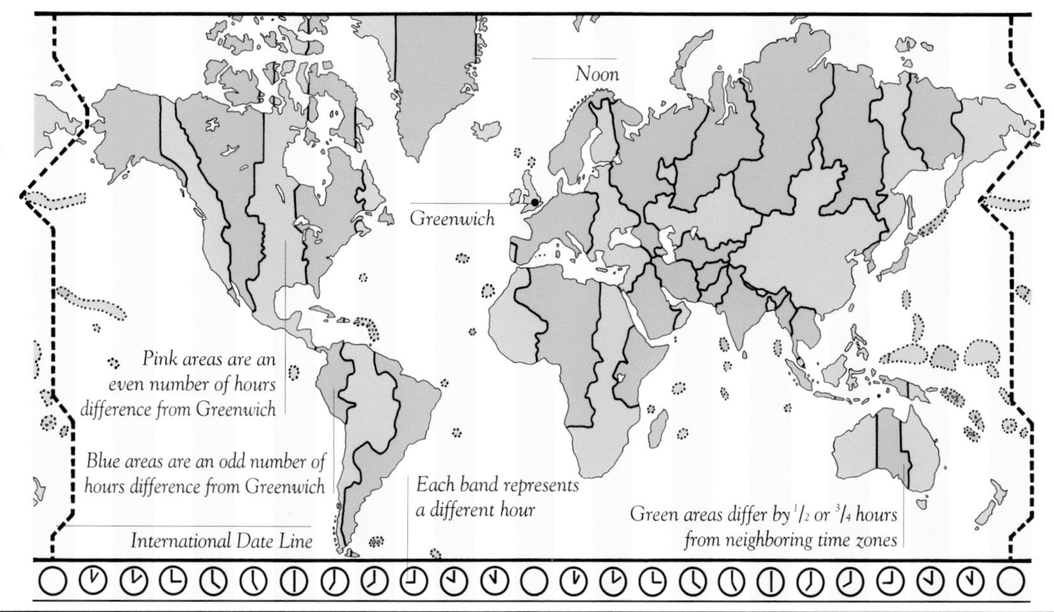

Greenwich mean time
In 1675 King Charles II established the Royal Observatory at Greenwich in England, to find ways of making navigation safer. The method finally decided upon was to compare a ship's time at sea with Greenwich time (ships took clocks on board that were set to Greenwich time). In effect, this gave the ship's longitude, which when combined with its latitude gave the ship's position. In 1884 Greenwich time was adopted by an international conference as the standard to which all other times are compared.

The pin marks where you live

The side of the globe facing the lamp is in daytime

The side of the globe turned away from the lamp is in nighttime

Lost in time
Put a differently colored pin one-quarter of the way around the Earth from where you live. As the Earth turns, you will notice that it is noon first at one pin, then at the other. There are 24 hours in a day, so if the second pin is one-quarter of the way around the Earth from you, what is the time difference from the first pin? If it was halfway around the Earth, what would the difference be? To measure the distance in terms of longitude, see the information on time differences above.

No hiding place

The Global Positioning System (GPS) is based on signals from *Navstar* satellites, such as the one above, and can pinpoint your position anywhere on Earth. Radio receivers are tuned in to signals from the satellites, which reveal how far the receiver is from each satellite. By comparing measurements, GPS can tell the receiver's position.

The time of day

EVERY DAY the Sun appears to rise in the East, cross the sky, and set in the West. We see this day in and day out because the Earth is spinning on its axis at a steady rate, turning once in 24 hours. This regular spin means that we can use the position of the Sun in the sky as a clock to tell the time. The Sun is extremely bright, so the safest way to do this is by looking at the position of the shadow the Sun casts. Remember that many countries set their clocks ahead by an hour in the summer to take advantage of the extra hours of daylight (the clocks are moved back again in the fall). This is called Daylight Saving Time. An additional benefit of this system is that it helps people economize on their lighting bills.

EXPERIMENT

Finding north

In order to use a sundial, first you need to locate north. The easiest way to do this is to use a compass. Another method is to put a stick in the ground in an open space from which the Sun is visible. Place a large piece of paper on the ground at the base of the stick, and at noon (or 1 p.m. during Daylight Saving Time) mark the shadow of the stick. The direction in which the shadow points is due north (south, if you are in the Southern Hemisphere).

YOU WILL NEED
- *stick*
- *paper*
- *pen*

Mark where the shadow falls

The stick will cast a shadow

Sundials

Mechanical clocks were not invented until about A.D. 1200. Until then, people had to use other methods for telling the time. The earliest people probably just watched the movement of the Sun across the sky. Sundials use shadows cast by the Sun to measure the time. They come in many forms—some, like the one in Jaipur, India, are large buildings; others are small enough to be placed outside in a garden. This ivory sundial is a pocket-size version that was made in Germany in 1648. A compass indicates north. Because they were easy to carry around, sundials like this were very popular in the 16th and 17th centuries. They are the "ancestors" of the even smaller modern watch.

The shadow cast by this string indicates the time on the dial below

Compass indicates north

EXPERIMENT

Making a sundial

If it is a sunny day, you can tell the time by using a sundial. This is a simple sundial you can make with a jar, some paper, and a knitting needle. If you are in the Southern Hemisphere, substitute "south" for "north" in the following instructions.

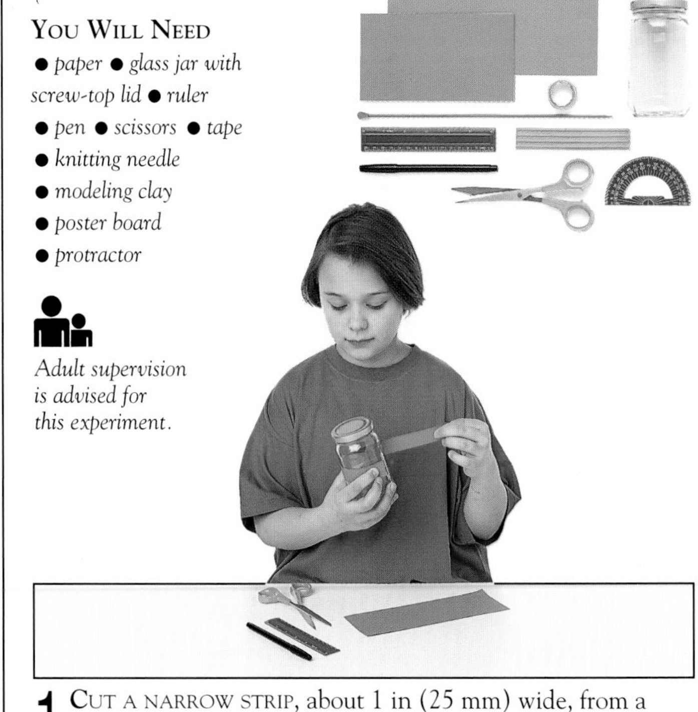

YOU WILL NEED
- *paper*
- *glass jar with screw-top lid*
- *ruler*
- *pen*
- *scissors*
- *tape*
- *knitting needle*
- *modeling clay*
- *poster board*
- *protractor*

Adult supervision is advised for this experiment.

1 CUT A NARROW STRIP, about 1 in (25 mm) wide, from a piece of paper. Wrap it around a glass jar, and trim the strip so that it fits the jar exactly. Do not fasten it to the jar.

2 UNWRAP THE PAPER, and measure it. Mark 24 equally spaced lines, and number them from 1 to 24 (from right to left [or left to right if you are in the Southern Hemisphere]).

The equation of time

Because of the Earth's orbit and tilted axis, sundial readings vary slightly from clock time. This chart shows how they differ through the year. Use it to correct your sundial readings.

(mins) J F M A M J J A S O N D (months)

20
15
10
5
0
5
10
15

Sundial ahead of clock

Sundial ahead of clock

The straight line shows clock time

Sundial behind clock

The wavy line shows the sundial reading

Sundial behind clock

3 WRAP THE PAPER around the outside of the jar, about halfway up the sides. Cover the paper with tape to fasten it to the jar and prevent it from getting wet if it rains.

4 PIERCE THE LID of the jar with a knitting needle. Slide the needle through, and put some modeling clay on the end. Screw the lid back on to the jar.

5 MAKE A POSTER-BOARD BASE, with a right angle at the top and the angle beneath it equal to your latitude (p.18). Add a poster-board strip to the top to keep the jar from slipping.

6 OUTSIDE, on a clear day, align the base from north to south (see "Finding north," opposite), and place the jar on top of it so the "12" faces north (down the slope). The shadow of the knitting needle on the paper strip will tell you the time.

A shadow will fall against the paper

This end of the base points north

Ensure the clay is in the middle of the base of the jar

This angle is equal to your latitude

The time at night

IF YOU OBSERVE the stars on a clear night, you will see that they appear to move slowly across the sky—just as the Sun appears to move across the sky during the day. Like the Sun, the stars rise in the east and set in the west within a fixed period of time. This is because the Earth rotates on its axis at a steady rate. One star, however, never moves. This is Polaris—the star that lies almost directly in line with the North Pole. In effect, the Earth turns underneath this star, which makes it look from Earth as if all the other stars were circling Polaris in the sky. For centuries, people have used the wheeling stars, as well as the Sun, to measure the passing of time.

Photographing star trails
The motion of the stars across the sky at night makes patterns that can be recorded by camera. This picture shows stars in the Northern Hemisphere circling Polaris. To capture this on film, fix your camera on a tripod so that it points north (for the Northern Hemisphere) or south (for the Southern Hemisphere). Focus the lens at infinity, and open the shutter. Leave the camera undisturbed for anything from 3 minutes to several hours. When the photographs are developed, the stars will appear to be drawn out into trails. If there is moonlight or light pollution, the photographs will have a bright background that obscures the stars. This is called "sky fogging."

Finding Polaris

As we look at the sky from the Northern Hemisphere, Polaris seems to stay in the same place—due north—while the other stars revolve around it. Because it is so well known, many people expect Polaris (also called the Pole Star or the North Star) to be immediately obvious, but in fact it is not a very bright star. However, you can find it easily with the help of the stars of the Big Dipper (also called the Plow), which appear from Earth to revolve around Polaris.

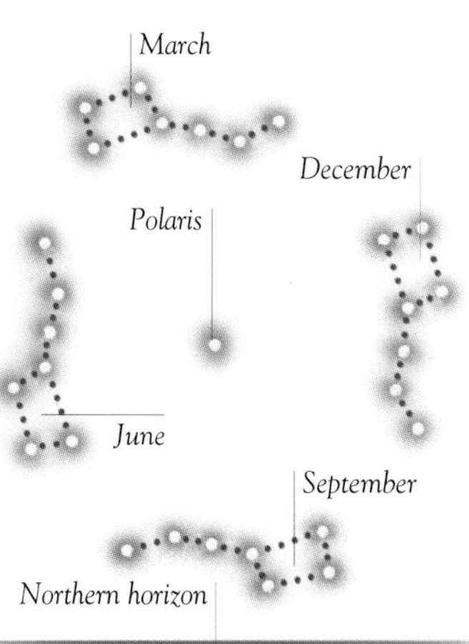

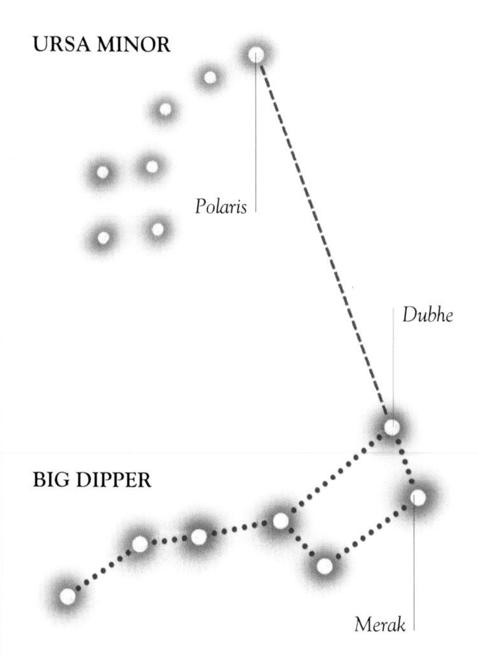

1 Find the Big Dipper
Look around the northern part of the sky until you find the Big Dipper (see above for its positions at midnight during certain months). Identify Dubhe and Merak (see left)—these are the pointers.

2 Mark the positions
Hold up a piece of poster board about 1 ft (30 cm) long, so that Merak is near one end. On the edge of the poster board, mark where Dubhe and Merak lie. Use a ruler to measure the distance between the marks. Make a third mark to the other side of Dubhe, 5.25 times the distance from Merak to Dubhe.

3 Find Polaris
Hold the poster board up to the sky again, with your original marks next to the pointers. Polaris should lie by the third mark.

Clock stars

The ancient Egyptians told the time at night with the aid of 36 "clock stars," which were equally spread across the sky. To do this, they first sighted south with a plumb line called a "merkhet." As each Egyptian night-hour passed (40 minutes in modern time), the next clock star passed the plumb line. In the summer, when the nights were short, only 12 clock stars passed the merkhet between dusk and dawn. The Egyptians began to refer to the "12 hours of night," and they matched them with 12 hours of daytime to make a total of 24 hours.

EXPERIMENT

Making a star-dial

As the Earth turns once every 24 hours, the stars seem to move around Polaris like the hands of a giant celestial clock. But there is one complication: the positions of the stars change during the course of the year. This must be taken into account if you are using the stars to tell the time at night. A star-dial is an instrument that, when adjusted to the correct date and pointed at Polaris, shows the correct time. Star-dials were common in medieval times.

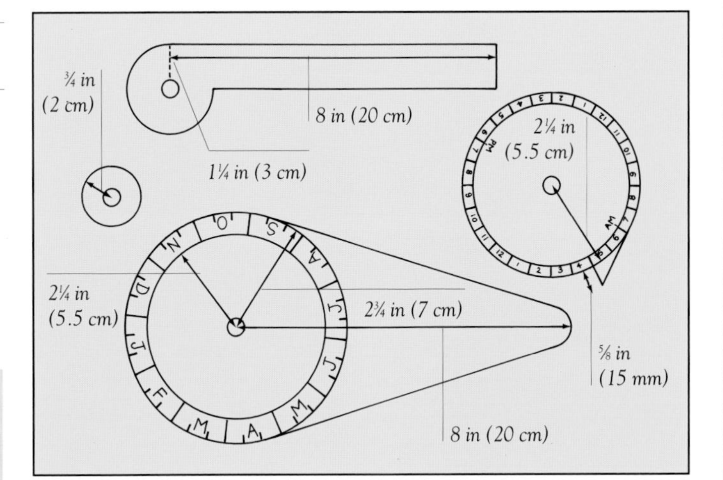

You Will Need

- *stiff poster board* ● *scissors* ● *pen* ● *ruler* ● *protractor* ● *short tube* ● *glue* ● *pencil* ● *plastic putty* ● *compass*

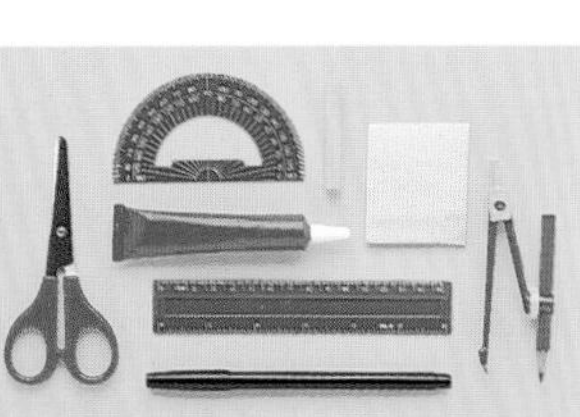

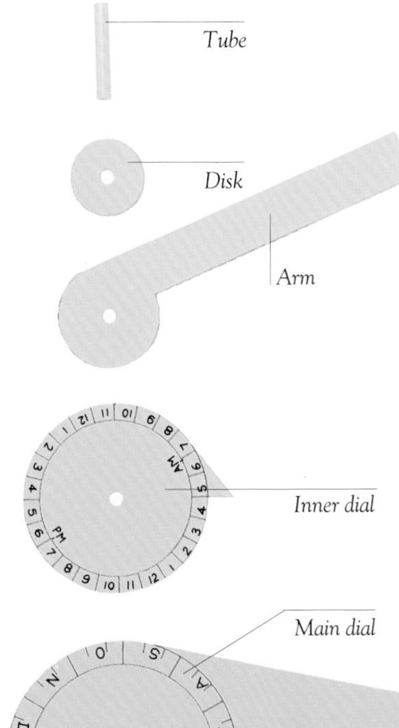

1 Cut out the main parts of the dial from stiff poster board (see the diagram above for the measurements). Write the months on the main dial and the hours on the inner dial, as shown.

2 Use a short tube, such as a drinking straw, to make a central pivot you can look through. Make sure the holes in the middle of the pieces of poster board fit tightly around the tube.

3 Assemble the pieces in the order above, and push the tube through the center. Put some glue on the back of the dial and the front of the small disk, so that when the glue dries the tube is stuck to the back and to the front disk. You should be able to turn the inner dial and the arm freely.

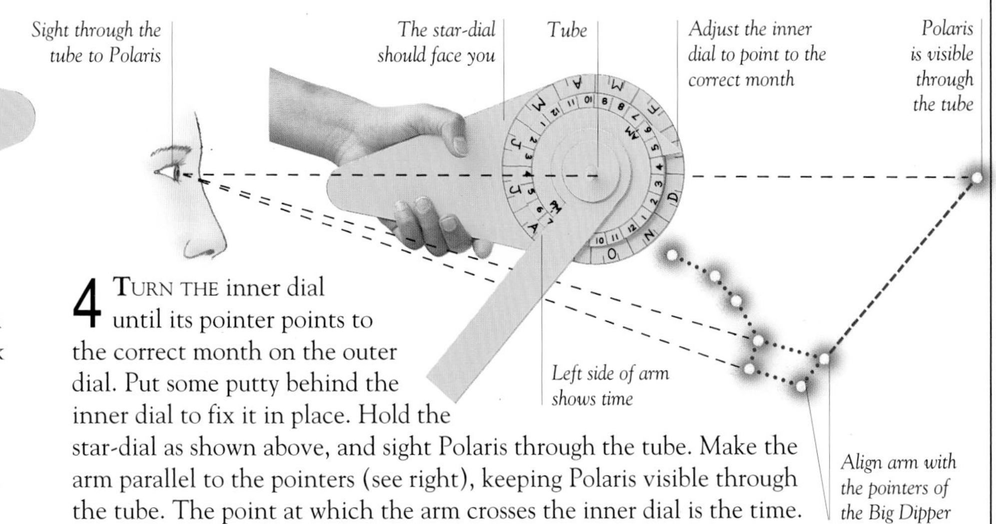

4 Turn the inner dial until its pointer points to the correct month on the outer dial. Put some putty behind the inner dial to fix it in place. Hold the star-dial as shown above, and sight Polaris through the tube. Make the arm parallel to the pointers (see right), keeping Polaris visible through the tube. The point at which the arm crosses the inner dial is the time.

Year after year

AS IT SPINS ON ITS AXIS, the Earth moves quickly through space following its orbit around the Sun—one orbit takes $365\frac{1}{4}$ days. One of the more obvious effects of this movement is the change in seasons—caused by the Earth's tilted axis, which dictates how much sunlight falls on different parts of our planet at different times. There are also other changes, such as the shifting star patterns in the night sky that repeat annually. Early astronomers noted that the Sun seems to move against the background of stars, an effect caused by our changing viewpoint. By watching the Sun and the stars, they could predict some seasonal events, such as the flooding of the Nile. Other people began to believe that the appearance of the night sky actually controlled events on Earth—the beginning of astrology.

EXPERIMENT

The seasons

The changing seasons have always been important because of the way they affect our environment and agriculture. This experiment shows you how the lengths of day and night change with the seasons. Put the lamp (the "Sun") on a table, and mark four positions around it about 3 ft (1 m) away, at right angles to each other: March, June, September, and December. Put the globe at each of these in turn. The axis should always point the same way—with the North Pole toward the "Sun" in June and away in December.

YOU WILL NEED
- *globe on a stand*
- *desk lamp*
- *plastic putty*
- *paper punch*
- *poster board*

EXPERIMENT

The band of the zodiac

As the Earth moves around the Sun, our perspective on the stars changes. In the Earth's orbit around the Sun, we see the stars only when we look away from the Sun because the view looking toward the Sun is blocked by its bright light. We know, but cannot see, that the Sun follows a particular path through the stars. This path is called the zodiac. In this experiment you can demonstrate the changing perspective.

YOU WILL NEED
- *compass* • *pen* • *tape*
- *flashlight bulb* • *wires and clips* • *battery* • *table-tennis ball*
- *wooden skewer* • *paintbrush*
- *foamcore* • *scissors*
- *paper clips* • *protractor*
- *paper* • *screwdriver* • *ruler*

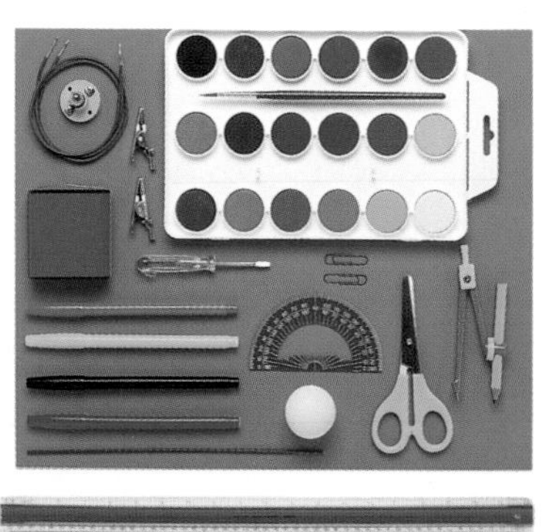

1 DRAW TWO circles on paper, one with a 10-in (25-cm) radius, the other with a 16-in (40-cm) radius. Cut out the larger circle, and mark six equally spaced lines across its diameter. Label the 12 "pie" sections with the months in a counterclockwise direction.

2 PUSH A wooden skewer through the table-tennis ball to make an axis. This represents the Earth. Push the end of the skewer into the foamcore, so that the "Earth" is at an angle of 23° to the vertical. Stabilize the "Earth" with another piece of foamcore.

The zodiac
Use the diagram below to copy the constellations.

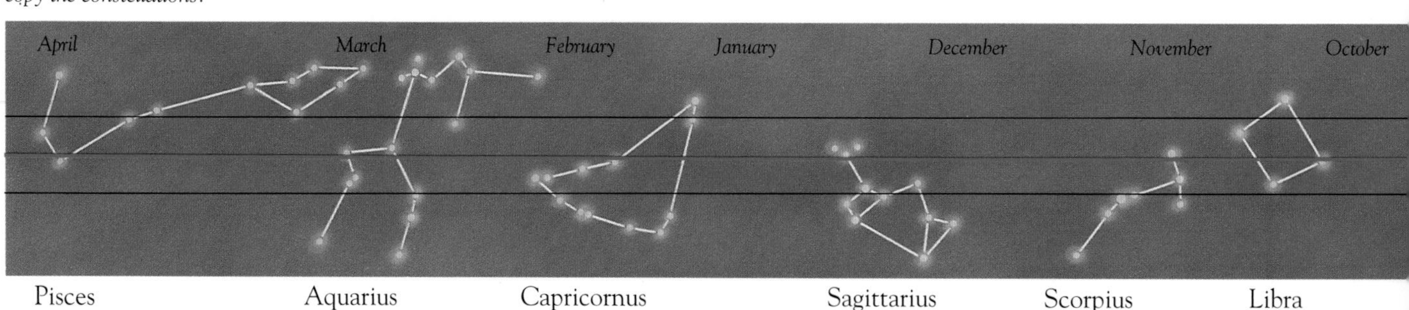

1 FIRST, INVESTIGATE day and night changes with the seasons. Align the top of the globe with the lamp's light bulb. Make five blobs of plastic putty, and stick them on the globe at the Equator, the poles, and halfway between the two. Place the globe at the December position, with the South Pole toward the Sun.

2 WITH THE LAMP ON, slowly turn the globe through one day (moving counterclockwise as seen from above). How much daylight do the North Pole and the South Pole get? Repeat the experiment at the March, June, and September positions to see how daylight changes with the seasons.

Temperature changes

Punch two holes in poster board, one above the other. Shine a light through the holes at the globe, in the December position (see left), with one patch of light on the Equator and the other on the Southern Hemisphere. Compare their brightnesses: the more intense the light, the higher the temperature on Earth. Now put the globe in the June position and see what is different.

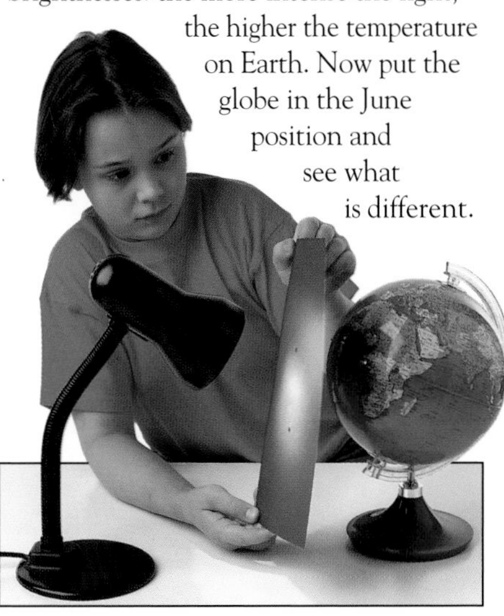

3 DRAW THE ZODIAC on paper 32 x 4 in (80 x 10 cm), copying it from the diagram below. Add the months and the red line (the apparent path of the Sun). The Moon and planets would keep within the black lines. Join the ends of the paper to make a circular strip, and place this over the large circle. Connect a flashlight bulb (the "Sun") to a battery, and put it in the center on a foamcore support.

4 SET "EARTH" ON the base in the June position, with the axis pointing toward the "Sun." Keeping the axis pointing in the same direction in space, move the "Earth" through the months. Note at each point the constellation behind the "Sun." You will see the "Sun" moving through the zodiac as the "Earth" moves.

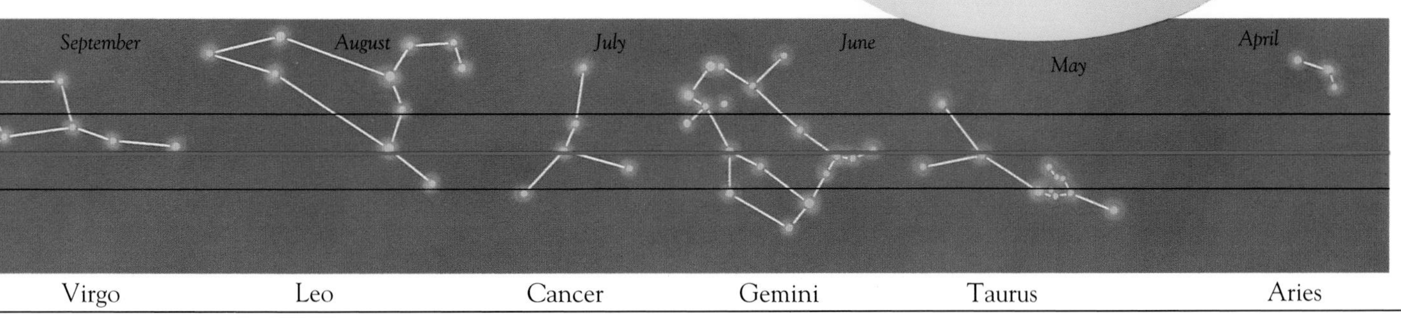

Our window into space

LOOKING OUT INTO SPACE is like looking at a beautiful landscape through a dirty window. Earth's atmosphere—although it is quite thin—gets in the way of our view of the Universe and often distorts and obscures what we see. One obvious effect of having an atmosphere is that it blots out the stars during the day. It is because of the atmosphere that the sky is blue and sunsets are red. The atmosphere can also produce some strange optical effects. A "blue Moon," for instance, is caused not by a change in the Moon, but by dust in the Earth's atmosphere. The appearance of halos around the Sun or the Moon is caused by high-altitude ice crystals bending the light away from its usual course. Halos may be a warning of bad weather.

Lunar halos

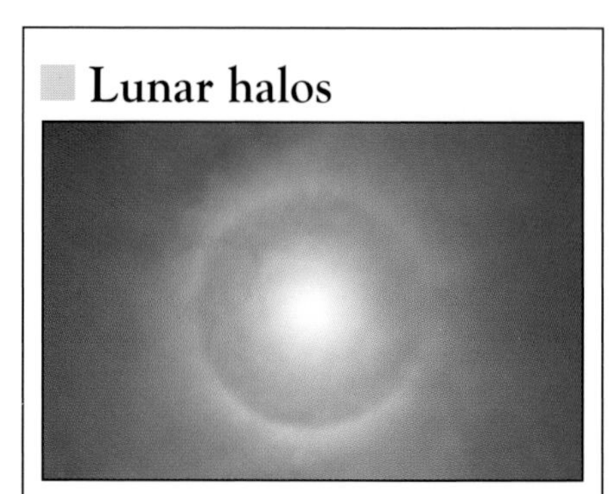

Several times a year, you may notice a bright ring around the Moon—a halo. This has nothing to do with the Moon itself. Instead, it is caused when light reflected toward the Earth by the Moon is refracted, or bent, inside tiny crystals of ice high in the Earth's atmosphere. Sunlight can also be refracted like this, making a halo appear around the Sun. Often, the halo is especially bright at two points, one on either side of the Sun, and these bright patches have faint rainbow colors. They are called parhelia. If the real Sun is dimmed by clouds, the parhelia can look as bright as the Sun itself, so it looks as though there were three Suns in the sky.

EXPERIMENT

Why is the sky blue and the sunset red?

White sunlight consists of a range of colors (pp.96–97), but we can divide it roughly into blue and red shades. Clumps of air molecules in the atmosphere scatter blue light, but not red, making a blue sky. As the Sun sets, its light must pass through more and more of the atmosphere, losing its blue part and appearing red. Create your own "atmosphere" and "sunlight."

Adult supervision is advised for this experiment – NEVER mix bleach and ammonia.

YOU WILL NEED
- *1-gallon (5-liter) clear container*
- *household disinfectant*
- *teaspoon* ● *flashlight*
- *white poster board*
- *water*

1 FILL THE CONTAINER with water. Add 6 teaspoons of disinfectant. The solution looks clear, but it is now filled with tiny particles. Shake well.

2 SHINE A FLASHLIGHT through the container (the "atmosphere") onto a poster-board screen. What color is scattered? What color reaches the screen?

3 TO SEE DAYTIME "sunlight," pour away half the solution. Ask an adult to hold the container up, and shine a flashlight down through it. What color is the "sky"?

EXPERIMENT

Why we cannot see the stars during the day

The stars are above us during the day, even though we cannot see them. Our view is blotted out because bright sunlight is scattered by the atmosphere and overpowers the faint stars. You can make your own "stars" and "atmosphere" out of tracing paper and acrylic plastic.

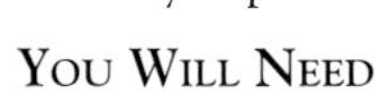

YOU WILL NEED

- *desk lamp*
- *stick-on stars*
- *dark poster board*
- *acrylic plastic*
- *tracing paper*
- *tape*

1 STICK stars on the poster board, and prop it up vertically. Cover the plastic with tracing paper, and put it 6 in (15 cm) in front of the stars.

2 PLACE THE lamp ("Sun") between the stars and the plastic ("atmosphere"). Point the lamp at the stars. Turn it on to see the stars through the "atmosphere." Now turn the lamp toward you. Can you see the stars?

4 TO MAKE A "SUNSET," gradually bring the "Sun" down. Ask a friend to watch, looking through the container and slowly moving upward as the "Sun" goes down. As your friend looks through thicker and thicker layers of "atmosphere," the light loses more and more blue. What colors appear, and in what order, as the "Sun" sets? Change places with your friend, and try it again.

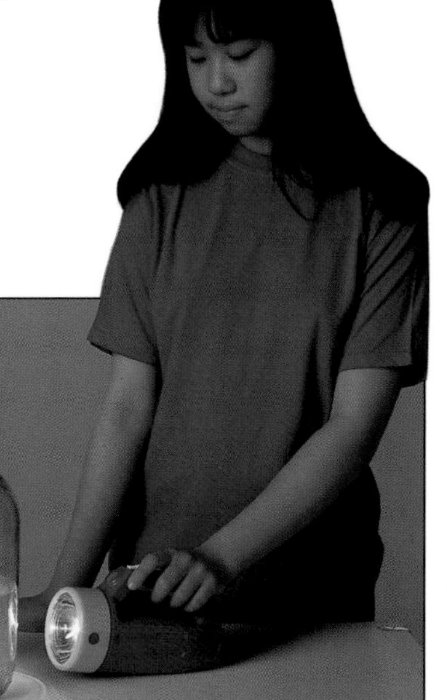

The atmosphere

The Earth's atmosphere is about 125 miles (200 km) high. It is densest at the bottom, gradually thinning out into space. Most clouds lie in the bottom layer of the atmosphere, the troposphere, which reaches to about 6 miles (10 km). Above that is the stratosphere, where there is a layer of ozone, a substance that protects us from the Sun's damaging ultraviolet radiation. As holes appear in the ozone layer (due to pollution), some of this radiation can get through and damage our skin and eyes. The atmosphere spoils our view of the Universe, so the best sites for Earth-based observatories are on top of tall mountains. The largest telescope on Earth, the Keck Telescope, is located on a mountain peak in Hawaii. Even more effective is the Hubble Space Telescope, which has a viewpoint above the atmosphere, floating in space.

***Ionosphere**: the region in which the spectacular auroras occur*

***Thermosphere**: the hottest region, where much of the Sun's radiation is absorbed*

***Mesosphere**: a region higher than 30 miles (50 km) and cooler than the regions above*

***Ozone layer**: a protective region in the lower part of the stratosphere*

***Stratosphere**: a region extending up to 30 miles (50 km) above sea level, with volcanic dust eruptions circulating in it*

***Troposphere**: the region closest to Earth; it has clouds and rain*

Rocket

Aircraft

EARTH

Light pollution

THE STARS ARE a beautiful sight and have been an inspiration not only to astronomers but also to poets, painters, and philosophers throughout the centuries. But today, although the stars are as bright as ever, light pollution is threatening our view of the dark night sky—eventually, perhaps, the only place we will be able to see stars will be in the artificial sky of a planetarium. Light pollution often appears as a dull orange glow in the night sky. It is caused by light from roads, towns, and cities reflecting off particles of dust and dirt in the atmosphere. Because of light pollution, only the brightest stars can be seen from most cities and often it is nearly as bad in the countryside. Some major observatories have already been forced to close down because of light pollution.

EXPERIMENT

Streetlight pollution

Most light pollution comes from streetlights. As a result, the sky in a city is never completely dark. You can make models of various streetlights to see which pollute the most, and then you can try designing a better streetlight.

Adult supervision is advised for this experiment.

YOU WILL NEED

- *empty toilet-paper roll*
- *3 small flashlight bulbs with holders*
- *clips and wires*
- *pliers* ● *4.5-V battery* ● *modeling clay* ● *glue* ● *wooden skewers* ● *poster board*
- *wire coat hanger*
- *scissors* ● *foamcore*
- *tape* ● *wire cutters*

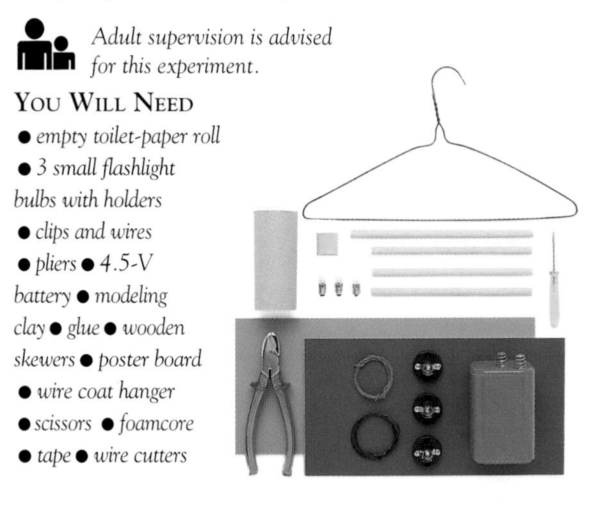

1 MAKE A BOX with an open front out of foamcore (see the diagram at right for measurements). Divide it into three equal-size compartments, each about 1 ft (30 cm) wide. Use glue to attach all the sections. Next make three streetlight models. Each consists of a flashlight bulb on a stalk. The first has a bare bulb, and the other two have partial hoods over the bulbs to prevent light pollution.

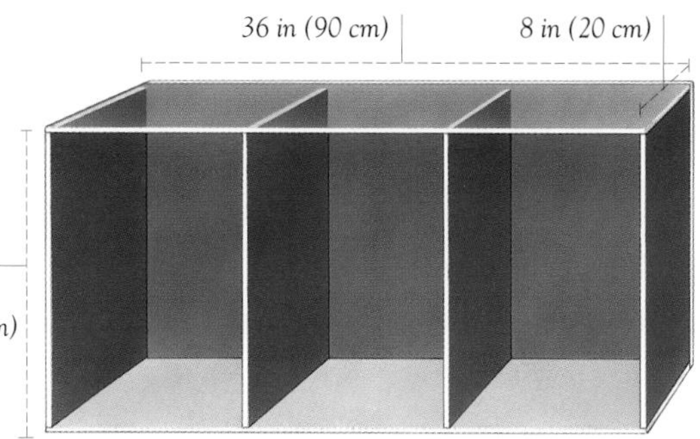

Measurements for the box

The display box for the streetlight models measures 16 x 36 x 8 in (40 x 90 x 20 cm), and it has one open side. The inside is subdivided into three compartments, each measuring 16 x 12 x 8 in (40 x 30 x 20 cm).

2 TO MAKE THE STREETLIGHTS, ask an adult to cut two 6-in (15-cm) pieces of coat hanger and to bend them into identical question-mark shapes. These will hold the hoods.

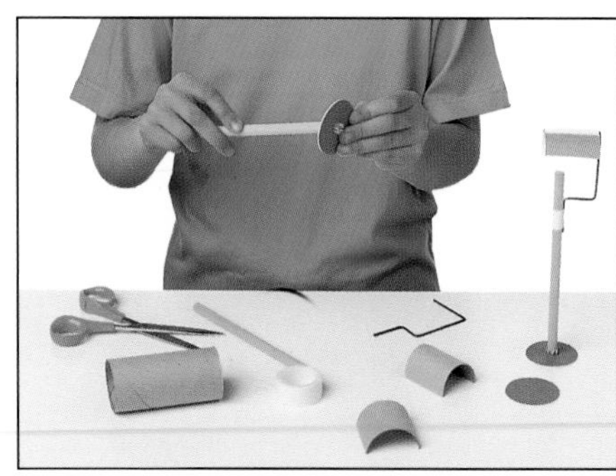

3 MAKE TWO curved hoods out of a toilet-paper roll—one, half the curve of the roll; one, a third. Fit the skewers into poster-board circles for bases, and tape the hoods to the wires and skewers.

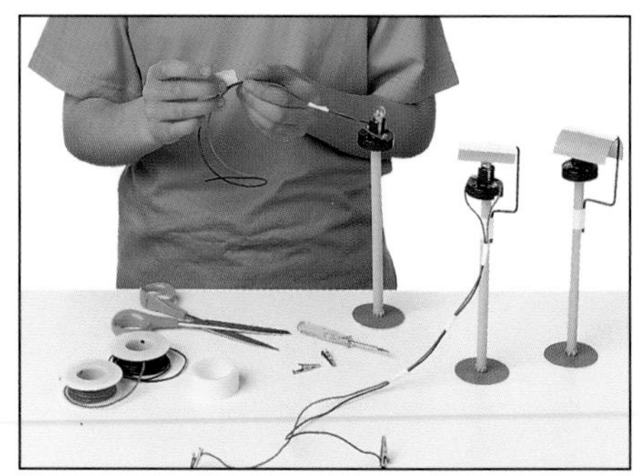

4 PLACE THE BULBS in their holders on the skewers with clay, and attach the wires and clips to the bulbs. Tape on the hoods as above, and place one streetlight in each compartment.

Light pollution survey

You can monitor light pollution by carrying out a survey in your local area. The constellation Orion is visible in both the Northern and Southern Hemispheres from November to May. Trace copies of the stars of Orion, and give a copy to each person taking part in the survey. Try to include people who live in different parts of the city and the country. Ask the participants to go outside on a clear night, let their eyes adjust to the darkness, and count how many of the stars in the diagram they can see. Severe light pollution will hide the fainter stars. People living under the bright lights of a city may see only the brightest stars that make up the outline of the constellation, while people in a darker area may be able to make out more than 20 stars. Find out how many stars each participant was able to see. Then make a map of the whole area, and write down the number of stars visible at each location.

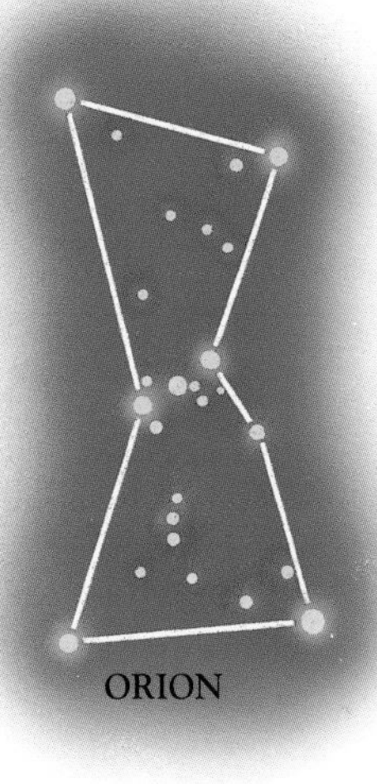

Drowned by light

Extensive light pollution has become a problem only in this century. Some older observatories that were built years ago were positioned in areas far enough from cities so as not to be affected by light pollution. But some cities grew so big that they overtook the observatories, making them useless. The Meudon Observatory in Paris, France (above), has suffered so badly from light pollution that it cannot be used anymore. Nowadays, strict rules control the lighting around many observatories, such as Kitt Peak in Arizona and La Palma in the Canary Islands. The best site would be the far side of the Moon.

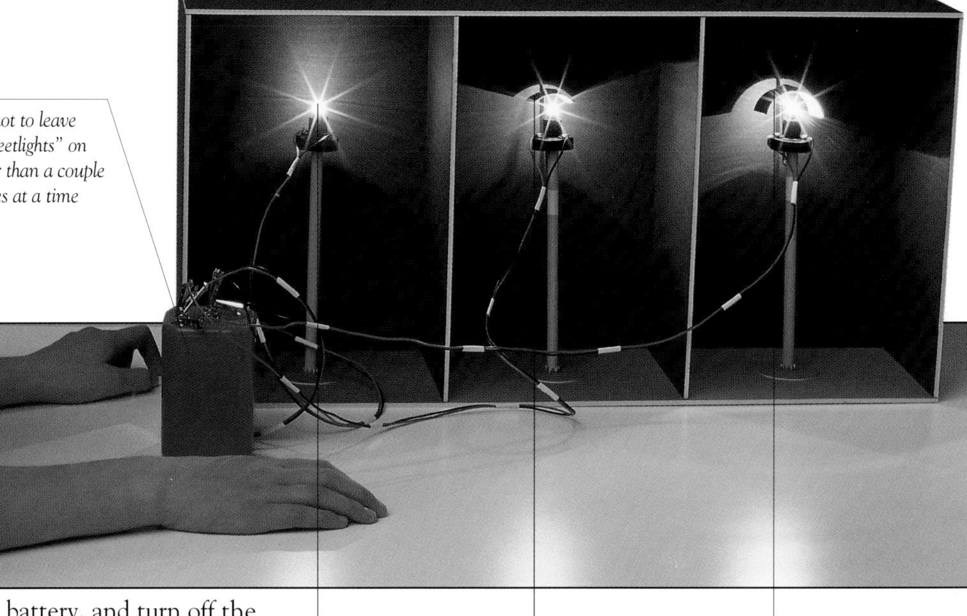

Be sure not to leave your "streetlights" on for longer than a couple of minutes at a time

No hood covers this light, so much of the light shines upward

This hood prevents only a small portion of the light from polluting the sky

The fuller hood still allows some light to escape into the sky

5 CONNECT ALL THE WIRES to the battery, and turn off the lights in the room. How much "spare light" spills out of the three lights? Which hood is best? Try making other hood shapes to see which allows the least light pollution.

Breaking free

IN THE 20TH century we have made a great leap forward in our exploration of the Universe: we have managed to break free of our planet and travel in space. In fact, space isn't very far away. If you had a car that could go straight up and the Earth had no gravity, space would be only a 2-hour drive. The speed needed to overcome the enormous pull of the Earth's gravity—known as the escape velocity—is 7 miles/sec (11 km/sec), or nearly 25,000 mph (40,000 km/h). Rockets can travel at these enormous speeds, and unlike aircraft, they work perfectly in the airless environment of space. A rocket burns its fuel inside a combustion chamber. The gases expand and rush out through the rocket's exhaust, giving the push that drives the rocket forward. Rockets serve as "launch vehicles," sending space probes on trips through interplanetary space or placing satellites into their correct Earth orbits. Today, hundreds of satellites orbit our world, each doing a different job.

The rocket pioneers

Early in the 20th century, a few pioneers experimented with rockets as a means of getting into space. Many people thought they were crazy, but they were simply way ahead of their time. The Russian Konstantin Tsiolkovsky, for instance, designed a multistage rocket (like those used in the *Apollo* space program) before the Wright brothers' plane even got off the ground. Other important pioneers were Wernher von Braun and Robert Hutchings Goddard.

Konstantin Tsiolkovsky
This teacher was writing articles about space flight as early as 1883.

Robert Hutchings Goddard
In 1926 Goddard built the world's first liquid-fueled rocket.

Wernher von Braun
Von Braun designed the world's largest-ever rocket—the Saturn V moon-launcher.

EXPERIMENT

Which orbit?

A satellite circling the Earth closely (for example, a spy satellite) has to travel fast to avoid being pulled in by the Earth's gravity. Satellites in higher orbits (ones used to look out into space, for example) can travel more slowly. Some must orbit at exactly the same speed as the Earth turns, in order to "hover" over one spot and beam signals down. Here you can make your own "satellite."

YOU WILL NEED
- *string*
- *scissors*
- *eraser*

1 CUT A PIECE of string about 20 in (50 cm) long, and tie an eraser (the "satellite") securely to one end. Start swinging the eraser around in a circle, with the string at its longest.

2 MAKE SEVERAL MORE CIRCLES, shortening the string each time. Notice how much faster the eraser circles. Try to swing it around slowly with a short string—your "satellite" will not go into orbit because its speed is not enough to balance the gravity of the Earth.

EXPERIMENT

Building a rocket

A rocket's thrust is provided by gases bursting out through an exhaust opening at its base—just like air rushing out of a balloon. You can build your own "rocket" using a balloon.

You Will Need

- *ruler* • *scissors*
- *drinking straw*
- *long, thin balloon*
- *balloon pump*
- *clamp* • *thread*
- *tape* • *flower pot*
- *stones*

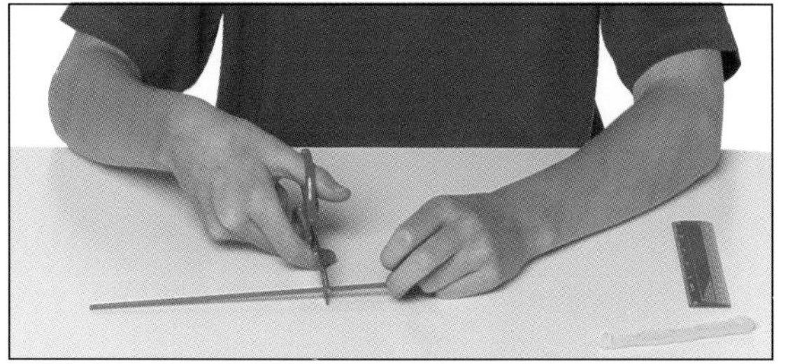

1 Cut two lengths of drinking straw, each 2 in (5 cm) long. Blow up the balloon (you may need the help of a pump), and fasten the end with the clamp.

2 String the thread through the pieces of straw. Attach the free end of the thread to the flower pot with some tape. If the pot is not very heavy, put some stones in the bottom to act as a weight.

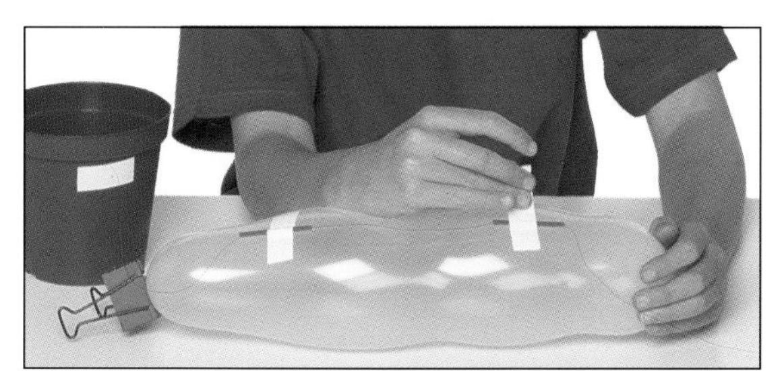

3 Tape the straws to the balloon as shown above. Unreel the thread, take it up to a high point, such as the top of a door frame or stairway, and fix it there. This is the flight path.

Rockets past, present, and future

Rockets, invented by the Chinese, have been used as weapons for thousands of years. The German V2 war rocket developed into the Saturn V moon-launcher. The space shuttle is the world's first reusable vehicle; eventually, space vehicles may take off like planes and switch to rocket mode in orbit.

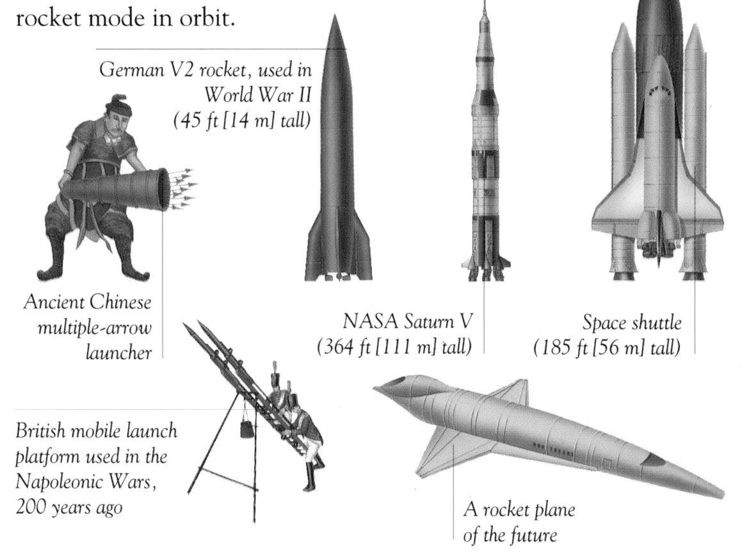

German V2 rocket, used in World War II (45 ft [14 m] tall)

Ancient Chinese multiple-arrow launcher

NASA Saturn V (364 ft [111 m] tall)

Space shuttle (185 ft [56 m] tall)

British mobile launch platform used in the Napoleonic Wars, 200 years ago

A rocket plane of the future

4 Position the pot so that the thread slopes upward. Remove the clamp and hold the balloon tightly—then release, and watch the "rocket" fly up the thread.

Space laboratories

SEVERAL TIMES EACH MONTH, spacecraft with people aboard fly over your head, hundreds of miles up. They may be cosmonauts living on a Russian space station or astronauts flying the space shuttle. A spacecraft in orbit around the Earth is a unique laboratory for scientific investigations. Scientists in orbit can study how objects behave when they have no weight, and they can view the whole Earth below and see the distant Universe clearly. Many tasks are carried out by the hundreds of satellites orbiting the Earth without people on board. Some satellites transmit telephone messages around the world and beam television programs into our homes.

Space stations
The Russian Mir *space station has been in orbit around the Earth since 1986. It has been occupied by cosmonauts from many countries, living and working in space—sometimes for over a year.* Mir *is made of interlocking modules, ferried into space as necessity and finances permit.* NASA *is planning a new space station to be used for manufacturing processes that require the unique environment of space.*

EXPERIMENT

Weightlessness in space

All objects in orbit fall toward the Earth at the same speed. They travel very quickly sideways, following a curved orbit, so they miss the Earth's surface. In principle, these objects are weightless because they are in "free fall." You can make your own "space capsule" to show this.

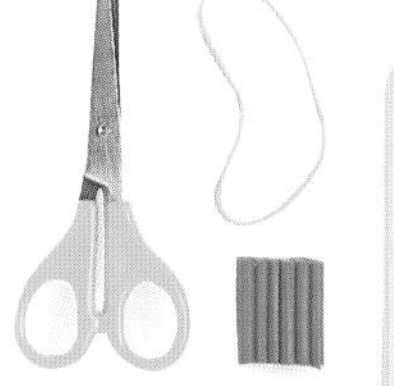

YOU WILL NEED
- *clear plastic bottle with screw-on cap*
- *scissors*
- *large rubber band*
- *plastic putty*

1 CUT THE RUBBER BAND to make a long strand. Screw the cap on the bottle over the rubber band, so that the band hangs slack from the cap.

2 REMOVE THE RUBBER BAND, and put a blob of putty on one end. Put it back, and note the effect of the weight of the putty on the band.

3 FROM A height of 1 yd (1 m), drop the bottle. What happens to the rubber band as the bottle drops? Does the band seem to feel any pull from the weight of the putty? The same thing happens to objects—including astronauts—inside an orbiting spacecraft that is in free fall.

Satellite spotting

There are hundreds of satellites in orbit around the Earth. They are dedicated space laboratories, equipped for tasks ranging from monitoring the Earth's weather to studying distant galaxies. Many people are surprised that you can see satellites crossing the sky, but metal satellites and their huge solar panels reflect sunlight very well. Some of them are so large that they outshine the brightest stars. The best time to view satellites is up to a couple of hours after sunset or before sunrise, when they can still catch the rays of the Sun. You don't need any equipment to do this—just let your eyes adapt to the darkness and watch the sky.

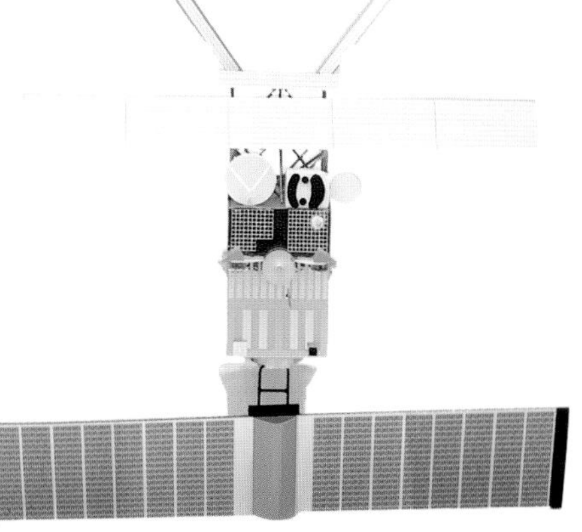

ERS-1
Launched into an orbit that takes it over the poles, Europe's ERS-1 satellite monitors the Earth's oceans as the planet spins underneath.

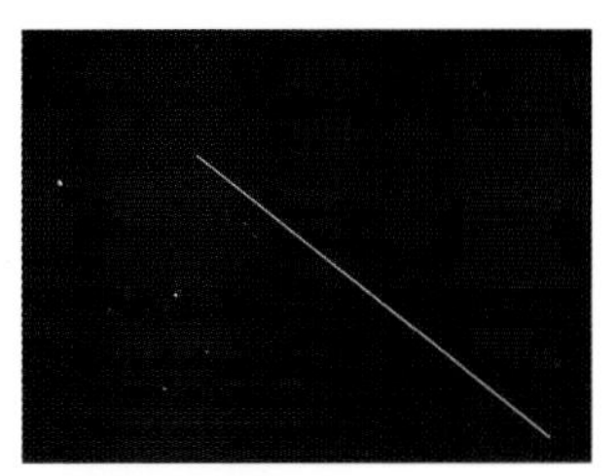

Satellite trail
Satellites usually appear as a uniformly bright light moving steadily across the sky. Although the sky is dark, they are high up in space and catch the Sun's rays.

An eclipsing satellite
If you see a bright light moving across the sky that gradually begins to fade (as above), it is probably a satellite moving from sunlight into the Earth's shadow.

European Retrievable Carrier
EURECA is a huge satellite. Launched into an east-west orbit, it stays in space for months at a time doing experiments.

Tumbling satellite
Short flashes of light moving in a straight line are the sign of a satellite that is tumbling along its orbit. Some spin deliberately; others are out of control.

Space experiments

Because everything in orbit is weightless, scientists can investigate how things would behave without the pull of gravity. They have measured how crystals grow, how chemical reactions work, how spiders spin webs, and how human blood cells grow. This research will lead to better computer chips and purer drugs to fight disease.

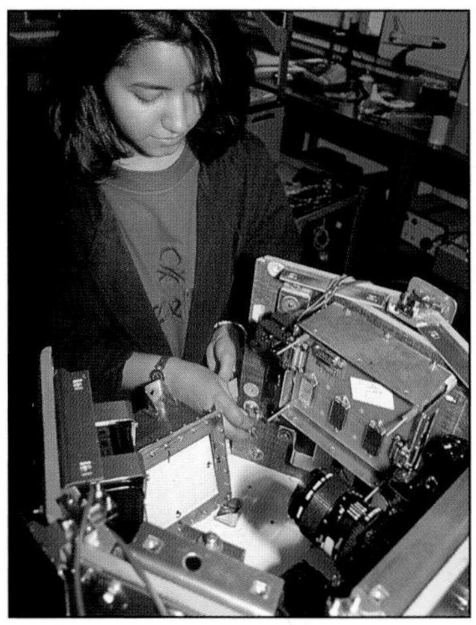

Preparing a chemical garden
A school pupil prepares a "chemical garden" to be carried on the space shuttle Endeavour.

Chemical stalks in solution
On Earth, the stalks grow up—like a plant—but in the weightless conditions of space the stalks grow sideways.

Living in space

Russians V. Volkov and S. Krikalev perform here for their fellow cosmonauts aboard the *Mir* space station. Living in the weightless environment of space means making many adjustments. For instance, you must take vigorous exercise to prevent your muscles from wasting away. Many people also become space-sick when they are first weightless.

The MOON

A barren world
Seen from the orbit of Apollo 12, the Moon (left) is a battered, desolate place with no water or protective atmosphere. By studying samples of moonrock (above) collected by the Apollo astronauts, scientists have learned about the Moon's structure and how it was born.

THE EARTH'S COMPANION IN space, the Moon, is as different a place from our own world as can be imagined. Its crater-scarred surface, the result of intense meteorite bombardment in the past, has not changed for billions of years. The Moon is completely dead. Even so, astronauts will almost certainly return to the Moon during the 21st century, and scientists are already working on plans to build permanent colonies for human settlement.

EARTH'S SATELLITE

PEOPLE ALL OVER THE WORLD, even if they know nothing about astronomy, can recognize the Moon. It is the brightest object in the sky after the Sun. At night the Moon may provide sufficient light for us to see our way around and sometimes even enough for reading. Before artificial lighting, moonlight provided some safety during the night.

The light of the Moon is not constant. Its brightness and shape (its phase) change over the course of a month—a word that derives from "Moon." Our ancestors watched the changing shape, or phases, of the Moon and used them to draw up a calendar. In fact, the Moon has no light of its own and shines only because it reflects the Sun's light. Occasionally, a full Moon fades from sight and may disappear entirely for an hour or so. This is called an eclipse of the Moon, and it can be a frightening spectacle for people who do not understand astronomy. The Moon becomes dark during an eclipse because it moves into the Earth's shadow, so that the Sun is no longer shining on it.

Glowing red
A lunar eclipse occurs when the Moon enters the shadow of the Earth. The Moon is illuminated by red light that has passed through the Earth's atmosphere and been bent toward the Moon.

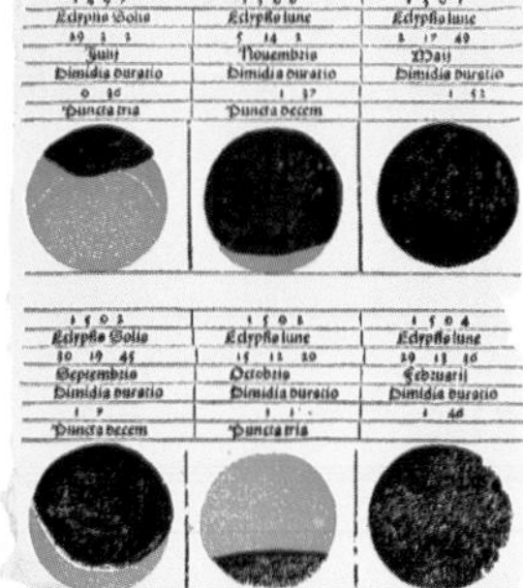

Columbus's calendar
This is the lunar calendar that saved the lives of the explorer Christopher Columbus and his crew. With its help he was able to predict an eclipse that astonished his captors. They freed Columbus and his men and looked after them until they were rescued.

Pulling force

In the 17th century Isaac Newton's questions about the Moon led to his theory of gravity. He knew that the Moon goes around the Earth in an almost circular orbit, and he wondered why it did not just wander off into space. As he idly gazed at his garden one fall day in 1665, he saw an apple fall from a tree. Newton suddenly realized that the force pulling the apple to the ground must reach way out into space and also hold the Moon in its orbit. He called this force "gravity" and proposed that every object in the Universe must have its own gravity. The Moon's gravity is weaker than the Earth's, but it does affect our planet. In particular, it pulls on our oceans, causing the tides.

The Moon is the only natural body to orbit the Earth. Some of the other planets have more than one moon. The record is held by Saturn, with at least 18 moons. Our Moon is one of the least interesting. One of Neptune's moons has eruptions that produce huge plumes of dark smoke, while a moon orbiting Jupiter has volcanoes that throw out hot sulfurous gases.

With most planets, the moons are much smaller than the planet itself. Our Moon is large in comparison—one-quarter the diameter of the Earth. It is actually bigger than the planet Pluto and almost as large as the planet Mercury.

Isaac Newton (1642–1727)
Newton was the physicist and mathematician who first explained the force of gravity.

Lunar landscape
This is the view from Apollo 11 *of the far side of the Moon. The rough, cratered terrain is typical of that area.*

For anyone starting out in astronomy, the Moon is a great place to begin. If you cannot see the Moon on your first night of observing, wait a couple of weeks, and it will be around again.

Lunar features

You can study the Moon using only the naked eye. As well as seeing the phases of the Moon, you can make out some markings on its surface—dark patches that make up the face of "the man in the Moon." The markings on the Moon are the same throughout the month: you always see the face of the man in the Moon, never the back of his head. Astronomers call the side of the Moon that faces the Earth the "near side." The part that is permanently turned away from the Earth is the "far side." If you lived on the near side of the Moon, you would always see the Earth hanging in the sky, but anyone living on the far side would never see the Earth at all.

On both the near and far sides, the Sun takes a month to go from sunrise to sunrise—a 2-week-long day and a 2-week-long night. We did not know what the far side looked like until the Soviet space probe *Luna 3* went around the Moon in 1959 and sent back photographs, showing a rough surface with fewer dark markings than the near side. (Sometimes the far side is called the "dark side of the Moon," but this is wrong: the far side is lit by the Sun in just the same way as the near side.)

The Moon is a dead world with no oceans, no atmosphere, and no vegetation. The dark markings are flat plains rimmed with high mountains—some higher than Mount Everest. Everywhere you look, there are craters: round basins that range in size from walled plains hundreds of miles wide right down to tiny pits that you can hardly make out with even the most powerful telescope.

Lunar origins

Although the Moon is so close, for many centuries astronomers were perplexed by it. They argued about the origin of the craters: Were they the remains of lunar volcanoes, or were they formed when meteorites crashed onto the surface of the Moon? More important, no one could decide on the origin of the Moon itself. Some scientists thought that it was originally a part of the Earth and that it whirled away from our planet because it was spinning too fast. Others said that the Moon was born somewhere else in the Solar System and that the Earth's gravity had grabbed the Moon as it passed by. We now have clues to the answers to many of the questions astronomers have asked over the centuries, thanks to the 12 *Apollo* astronauts who walked on the Moon's surface.

Spewing volcano
Before the advances of this century, some scientists thought craters on the Moon were formed by lunar volcanoes, as shown in this 19th-century illustration.

Exploring the far side
The Soviet Luna 3 *space probe was the first to send back pictures of the far side of the Moon.*

The first person to step on the Moon, on July 20, 1969, was the American astronaut Neil Armstrong, with the words: "That's one small step for a man, one giant leap for mankind." The astronauts of the *Apollo* missions found a barren, grayish world with a black sky—one of them called it a "magnificent desolation." They brought back moonrocks from six different landing sites. On the last three Moon missions the astronauts drove an electrically propelled Lunar Roving Vehicle—or "moonbuggy"—to collect rocks several miles from where they had landed. Soviet robot spacecraft brought back smaller samples of lunar soil from three other regions of the Moon.

Geologists have analyzed these rocks to work out the history of the Moon. The most-accepted theory about the birth of the Moon involves a giant catastrophe. When the Earth was very young, a wayward planet may have smashed into it, splashing out incandescent rocks that later came together to form the Moon. As bits of rocky debris continued to fall on to its surface, they formed craters. The Moon never had volcanoes, but lava seeped from its interior to form the great dark plains that make up the face of the man in the Moon. It is a dead, airless, and battered world.

Return visits

No one could live on the Moon without protection. But in spite of this, scientists around the world are planning missions to return. This airless, dead world has permanently dark skies, so it is an ideal place for an observatory. And because the Moon is our nearest neighbor, it is the ideal launchpad for missions to the planets. Within the next generation, people plan to build a moonbase largely buried beneath the ground for protection. Exposed buildings will be constructed from concrete made from lunar soil. Robot miners will extract valuable metals that are common in moonrock, such as titanium. These will be used to build lightweight spacecraft for further space exploration. The Moon will become the staging area for the exploration of worlds beyond.

Finding the Moon
This is a moondial, a simple tool you can make to find the Moon in the sky *(p.47)*.

Custom car
Eugene Cernan operates the Lunar Roving Vehicle on the Moon in 1972. Cernan, of the Apollo *17 crew, was the last person on the Moon.*

The Earth and the Moon

FOR TWO BODIES that are so close together in space—the Earth and Moon are only 240,000 miles (385,000 km) apart—it is hard to imagine a pair that could be more different. The Moon is airless, lifeless, crater-scarred, waterless, and has wild temperature swings. The Earth is lush, temperate, almost overflowing with life, and has a protective atmosphere. The Earth is 3.7 times wider than the Moon, but even with its larger size it is surprisingly massive. This means that the Earth is denser than the Moon, so the two bodies must be very different inside. The greater mass of the Earth means it has a stronger gravitational pull. Because the Moon's gravity is so weak, astronauts weigh only one-sixth their Earth weight when they stand on the Moon.

The airless Moon

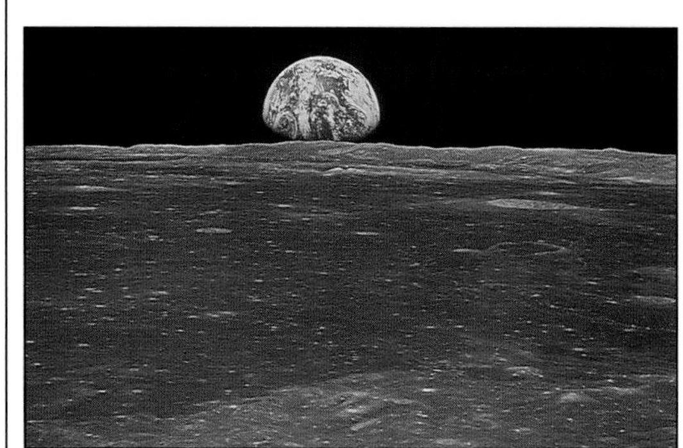

From the Moon, the view of the Earth and the rest of space is crystal clear. Because the Moon is relatively lightweight, compared with the Earth, its gravitational force is much smaller than the Earth's gravitational force. Lightweight, fast-moving gas molecules simply escape into space, so the Moon has no atmosphere. As a result, the Moon is exposed to all the extremes of space.

EXPERIMENT

Making a scale model of the Earth and the Moon

Loosely speaking, we say that the Moon travels around the Earth—but in fact they both move around a balance point at the center of gravity between the two. This point is the "barycenter" of the Earth-Moon system. The force of gravity depends on the masses of the objects and how far apart they are. Because the Earth is so much more massive than the Moon, the barycenter actually lies inside the Earth, just below the surface. The center of the Earth moves around this balance point. In this experiment you can make a scale model of the Earth-Moon system.

YOU WILL NEED
- *12 x 2 in (30 x 5 cm) poster board*
- *pen* • *pin*
- *scissors*
- *ruler*

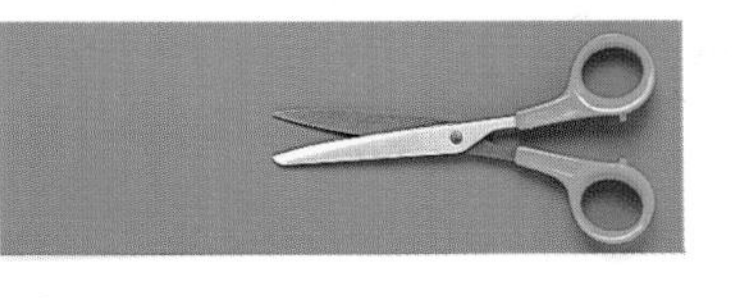

1 NEAR ONE END of the poster board, draw a circle ¼ in (6 mm) in diameter for Earth. For the Moon, draw a circle 1/16 in (2 mm) in diameter, 10 in (25 cm) from "Earth's" center.

2 MARK A DOT on the "Earth" ⅛ in (3 mm) from its center (within the circle), in the direction of the "Moon." This is the center of gravity. Stick a pin through the dot.

Moon

Earth

3 HOLD THE pin in one hand, and rotate the poster board around it in a counterclockwise direction. The "Moon" traces a large orbit—but observe the "Earth" as it swings around the balance point. It makes a small orbit around a point just inside the surface of the "Earth."

EXPERIMENT

Weighing the Earth and the Moon

The Earth has 81 times the mass of the Moon, so if you could put it on a giant cosmic scale it would be 81 times heavier than the Moon. If the Earth were made almost entirely of rock—like the Moon—then its volume would be 81 times larger, too. Below, you can make a model to show this. However, roughly half of the Earth's mass is contained in its iron core, which is denser and heavier than rock. At the end of this experiment, you can replace part of the "Earth" with an "iron core." This will show you why the Earth is such a heavy weight for its size.

YOU WILL NEED

- *3/4 in (20 mm) steel marble*
- *ruler*
- *scales*
- *plastic putty*

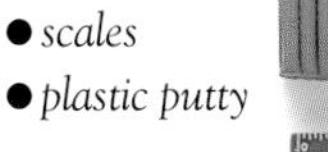

1 ROLL SOME plastic putty into a ball ½ in (1 cm) wide for the Moon. Now roll 81 balls of the same size. These are equal in weight to the Earth.

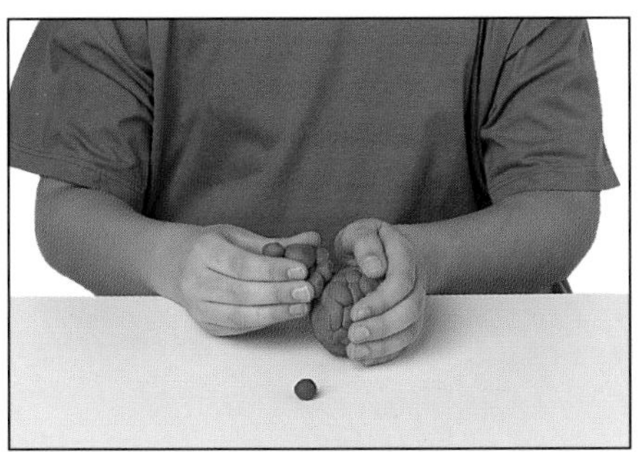

2 ROLL THE 81 BALLS together to make one large ball. Compare this "Earth" with your original "Moon"—is the "Earth" as large as you would expect?

3 NOW MEASURE the diameter of the two balls, and divide the "Earth's" diameter by the "Moon's." How much wider is your "Earth" than your "Moon"?

4 NOW, ROLL 41 balls and wrap them around a steel marble—the "Earth's core." The weight is the same as before, but what about the size?

Inside the Earth and the Moon

Although astronauts have drilled only a few feet into the Moon, we can investigate its interior by studying moonquakes and comparing them with the earthquakes on our planet. Because the Moon is a geologically dead world, moonquakes are much weaker than earthquakes and do not destroy terrain. By studying the way that waves from moonquakes pass through the Moon, scientists have discovered it is largely made of solid rock. Only near the center is the rock molten. If there is an iron core, it must be extremely small. This contrasts with the Earth, which has only a very thin layer of solid rock, the crust—the rest consists of molten rock and a large iron core.

Inside the Earth

Beneath the thin, solid crust—only 21 miles (35 km) deep—most of the Earth consists of a hot, rocky mantle. This molten rock can flow slowly up and down, like thick syrup. In the center of the Earth is a dense, heavy core made of molten iron.

Surface
Crust
Core
Mantle

Surface
Crust
Solid rock
Liquid rock
Possible core

Inside the Moon

The Moon has a thick crust that floated to the surface when the Moon was molten, soon after its birth. Now the interior rocks have cooled and solidified almost entirely. There may be a very small iron core, but scientists have not yet found evidence for it.

The Moon's orbit

THE MOON ORBITS the Earth about once a month. A sign of the Moon's movement is its changing phase—the amount of the Moon lit by sunlight as it swings around Earth. Each month the Moon grows from a new Moon, to a crescent, through to a full Moon, and then back again. These phases can be used as the basis for a lunar calendar, to measure the passing of time. If you observe the Moon closely, you will notice that the same side always faces the Earth. This is because the Moon rotates (turns on its axis) only once during each orbit. The only way to see the far side is from a spacecraft.

The Earth-Moon system

A "new" Moon is between us and the Sun: all the sunlight falls on the Moon's far side, so we cannot see it. As the Moon moves around the Earth, a little sunlight creeps around the edge, and we see a crescent. When the Moon is a quarter of the way around its orbit, we see half of it lit up—a phase called "first quarter." As it moves on, more and more of the Moon is lit up. When it is exactly opposite the Sun, all the sunlight falls on its Earth-turned face, and the Moon is full. Then the phases reverse: from full to "last quarter" and finally new Moon.

The inner illustrations show how the Moon is actually lit up—as if seen from space

Last quarter

Moon's orbit

Crescent

Sunlight comes from here

New Moon

Earth

Full Moon

The outer illustrations show the view from the Earth

First quarter

This bulging Moon is called gibbous

EXPERIMENT

The phases of the Moon

The Moon appears to change shape because of the way the Sun shines on it as the Moon circles the Earth. From the Earth, we never see the Moon's far side. (Sometimes this is wrongly called the "dark side" of the Moon. The far side in fact experiences sunshine and darkness just like the near side.) You might think that this means the Moon does not rotate at all, but it actually means that the Moon turns once in each orbit, as this experiment shows. In it you will view the "Moon" as if from Earth, while your friend views the "Earth-Moon system" from "space."

YOU WILL NEED
- *desk lamp*
- *old tennis ball*
- *pencil*
- *pen*

1 STICK A PENCIL into the tennis ball—the "Moon." Use a pen to make a dark spot on one side of the ball and a cross on the other side.

2 DARKEN THE ROOM, and stand in front of the lamp, holding the "Moon" so that the spot is facing you. Ask your friend to watch from nearby while you slowly spin around, keeping the "Moon" in front of you. You will see phases of the "Moon" as you turn. Ask your friend what the view from "space" is like. Ask each other this question: Is the "Moon" rotating, or am I always looking at the same side? Swap places, and try it again.

EXPERIMENT

Recording the phases of the Moon

You can see how long it takes the Moon to go around the Earth by observing the Moon's phases. However, there is a slight complication: during the month, the Earth moves on its circular path around the Sun. At the end of the month, the Sun is shining from a slightly different direction. This means the Moon has to overshoot its original position before the sunlight creates the same phase. So, we see the phases repeat after 29.5 days, but the actual period of the Moon's revolution around the Earth is 27.3 days. To create a Moon chart (below) you can make your observations with the naked eye, but it helps to use binoculars or a telescope. Look in a newspaper to find out what time the Moon is rising. The experiment takes nearly a month but won't be spoiled if you miss the odd night because of clouds.

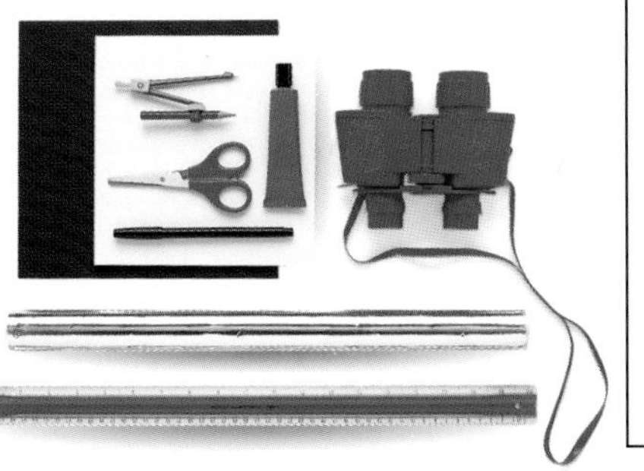

You Will Need

- *poster board*
- *black paper*
- *glue*
- *scissors*
- *aluminum foil*
- *compass*
- *thick black pen*
- *binoculars*
- *ruler*

Using lunar calendars

The period from one new Moon to the next is called the "lunar month." The problem with using lunar months as a calendar is that the year does not contain a whole number of lunar months. Twelve lunar months make 354 days, which is 11 days short of a year. So, over time, a particular month will change from being in the winter to being in the summer. Both the Islamic and Jewish cultures use lunar calendars. This building is crowned with the Islamic symbol of the crescent Moon.

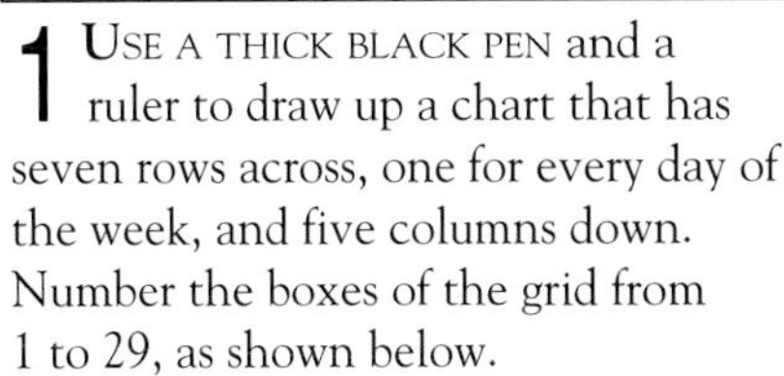

1 Use a thick black pen and a ruler to draw up a chart that has seven rows across, one for every day of the week, and five columns down. Number the boxes of the grid from 1 to 29, as shown below.

2 Draw a circle on black paper for each box, cut them out, and glue them down. Have your chart ready on a clear night to record the date and the Moon's appearance in Box 1. You can mark in the next 28 dates as well.

Your lunar calendar
This is what your chart will look like before you start recording the phases.

3 Every night, cut the Moon's shape out of aluminum foil and glue it down in the box for that day. If for any reason you cannot see the Moon, leave that night's box blank. Continue for several weeks, labeling the phases—first quarter, full moon, last quarter—as they appear. How long is it before the Moon looks the same as it did in Box 1?

Lunar eclipse

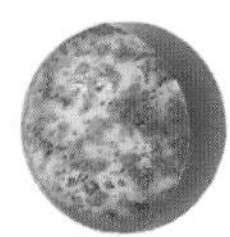

EVERY SO OFTEN we witness an alarming sight in the night sky. The full Moon is slowly engulfed by a black shadow until it disappears, though it may be visible as a strange reddish disk. This is a total eclipse of the Moon, and it occurs when the Moon moves out of the Sun's light and into the Earth's shadow. Eclipses do not happen every time there is a full Moon, because the Moon's orbit around the Earth is tilted in relation to the Earth's orbit around the Sun, and the Moon usually passes above or below the Earth's shadow. Newspapers and radio stations announce the details of forthcoming eclipses, but you can find out in advance by consulting an almanac. Make sure it refers to your part of the Earth: a lunar eclipse can be seen from only one-half of the world.

Columbus's eclipse

Eclipses can be predicted very precisely. In 1504 the explorer Christopher Columbus was marooned in Jamaica among islanders who became hostile after his men killed local people. By predicting an eclipse for February 29, Columbus frightened the islanders into feeding him and his men until they were rescued.

EXPERIMENT

Making an eclipse

Lunar eclipses occur when the Sun, Earth, and Moon are in perfect alignment. When they are out of line, there may be only a partial eclipse or none at all. You may have to wait a long time before an eclipse is visible in the sky where you live, but in the meantime you can make your own "lunar eclipse."

YOU WILL NEED
- *plastic putty*
- *wooden skewers*
- *small ball (about 1½ in [35 mm] wide)*
- *large ball (3 in [75 mm] wide)*
- *24 x 4 in (60 x 10 cm) piece of foamcore*
- *desk lamp*

1 PUSH A SKEWER into each ball. Put a lump of putty at each end of the foamcore, and push the skewers into each lump. The centers of the larger "Earth" and smaller "Moon" should be the same height.

2 SET UP a desk lamp ("Sun") 16 in (40 cm) away from "Earth," at the same height and shining directly at it. Pivot the foamcore, moving the "Moon" into "Earth's" shadow and out again to make an "eclipse." Raise the height of the "Moon." What happens?

EXPERIMENT

Why you can usually see the eclipsed Moon

The Moon's red color during a total eclipse is caused by the Earth's atmosphere, which bends, or refracts, sunlight shining on it and splits the light into separate colors. Only red rays survive the curved passage around the Earth, so reddish light illuminates the eclipsed Moon. You can bend white light to illuminate your own eclipsed "Moon."

YOU WILL NEED

- *clear-plastic bottle with top*
- *poster board*
- *ruler*
- *tape*
- *compass*
- *scissors*
- *plastic putty*
- *desk lamp*
- *milk*
- *water*
- *knitting needle*

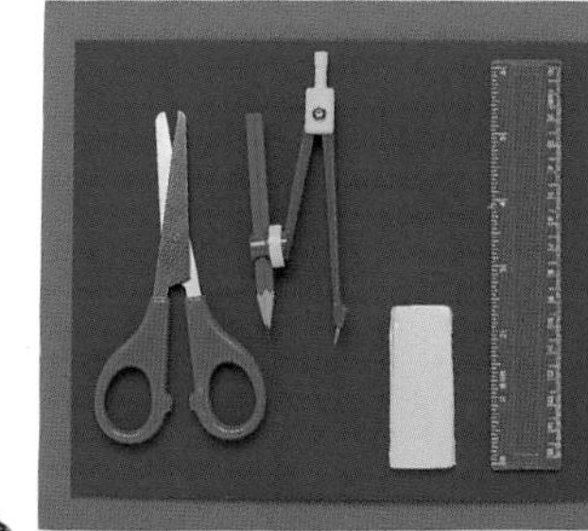

1 CUT A CIRCLE 4 in (10 cm) wide out of poster board. Tape it to the end of a knitting needle. Make a plastic putty base. This is the "Moon."

2 SHINE A LAMP at the "Moon." Darken the room. Raise some poster board ("Earth") between the "Moon" and the lamp until the "Moon" is in shadow.

3 IN THE BOTTLE mix 1 teaspoon milk with 1 cup water. Screw on the top. Put the "Earth" in position, and hold the bottle (the "atmosphere") sideways over it. This will bend the light. The "Moon" will be dimly illuminated—just as in an eclipse.

The Moon in darkness

Once every year or so, a lunar eclipse becomes visible on the night side of the Earth. During a total lunar eclipse, the Moon is covered completely by the darkest part of the Earth's shadow, the umbra, where no direct light from the Sun can fall. During a partial lunar eclipse, part of the Moon is in the umbra, and the rest of it is in the penumbra—an area of partial shadow, where some of the Sun's light has passed above and below the Earth.

Moon spotting

WE THINK OF THE MOON as lighting up the night sky. However, it can often be seen in the daytime—sometimes during the day it is easier to make out features on the Moon's surface because it glares less against the bright sky. In the past people depended on the Moon at night for illumination. At certain times of year it provided light that helped people to gather crops or hunt late at night. The Moon can sometimes have an unexpected appearance. When it is low in the sky, it can look extremely large. And when the Moon is a thin crescent, you can sometimes see the rest of its globe faintly illuminated—an effect called "the old Moon in the new Moon's arms."

The man in the Moon

What do you see when you look at the full Moon? Because we always see the same side of the Moon, we always see the same features—the Moon's many "seas," plains, and craters. With a little imagination you should be able to transform them into a face ("the man in the Moon"), a rabbit, or a woman wearing a necklace. The best time to observe the full Moon is when it is low in the sky, so that you see it through a greater thickness of the Earth's atmosphere. This helps cut down on glare, making the features easier to see.

EXPERIMENT

The old Moon in the new Moon's arms

When the Moon is in a crescent phase, you can sometimes see the rest of its globe dimly illuminated. The source lighting it is the Earth—a phenomenon called earthshine. Sunlight is reflected off the Earth's atmosphere and onto the Moon's surface, causing a faint glow. The cloudier the Earth's atmosphere, the more earthshine there is, and the brighter the Moon becomes. In this experiment you can create your own "earthshine." Watch the dark part of the "Moon" as you add "clouds" to a model Earth.

YOU WILL NEED

- *compass*
- *scissors*
- *poster board*
- *2 knitting needles*
- *tape*
- *glue*
- *small plastic ball*
- *plastic putty*
- *lamp*
- *ruler*
- *cotton balls*

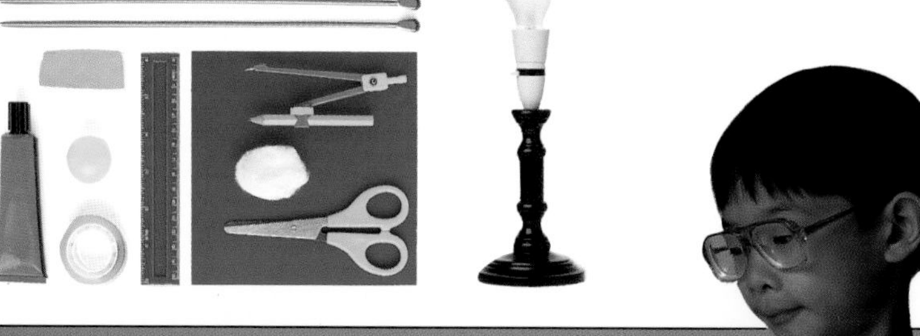

1 TAPE A CIRCLE OF poster board (the "Earth") measuring 5 in (12 cm) wide to a knitting needle. Stick the ball (the "Moon") on another needle.

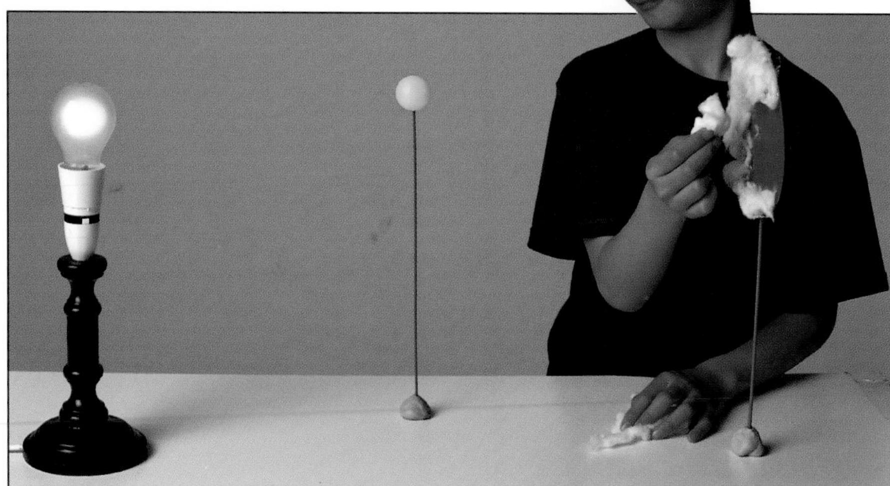

2 USE PUTTY TO stand the "Earth" about 2 ft (60 cm) from a lamp. Turn on the lamp, and turn out the lights in the room. Place the "Moon" between the lamp and "Earth," slightly to one side. Glue a cloudy cotton-ball "atmosphere" to the "Earth's" surface. What effect does this have on the "Moon"?

EXPERIMENT

Making a moondial

Even experienced astronomers can be taken by surprise by the Moon. Some nights you see the Moon; some nights you do not. During the course of a month, the Moon can appear in different parts of the sky, by day as well as by night. By making a moondial, you can find out—when the Sun is shining, at least—just where in the sky the Moon will be. The wedge-shaped base is made exactly as the base for the sundial (p.22).

** In the Southern Hemisphere, replace "counterclockwise" with "clockwise" and "north" with "south."*

You Will Need

- *poster board* • *compass* • *scissors* • *ruler*
- *pen* • *foamcore* • *glue* • *newspaper*
- *pushpin*

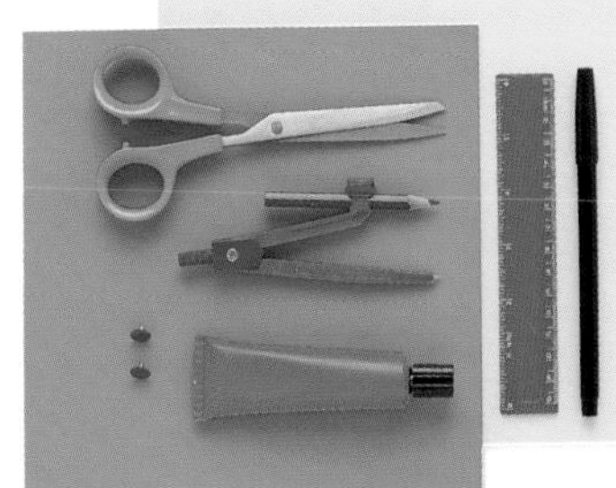

The Moon illusion

When the Moon is low on the horizon, it seems very large because our eyes compare it with objects such as trees and houses. Is this an illusion? Next time you see a full Moon rising, span its width with measuring tape held at arms length. Repeat this every hour for up to 6 hours. Is there any change in size?

1 Draw a circle about 8 in (20 cm) wide on poster board. Cut it out and draw another circle inside it with a diameter of 6 in (15 cm). This is the dial.

2 Divide the circle into 30 equal "pie" segments. Number them from 0 to 29 counterclockwise.* Mark the Moon's phases in Boxes 0, 7, 15, and 23, as below.

3 Make a wedge-shaped foamcore base, with a right angle at the top and the angle beneath it equal to your latitude.

4 Place the dial in the center of the base. Stick a pushpin through the center to secure it.

5 Find the Moon's phase from a newspaper, and put a mark in the corresponding box. In daytime, position the base outside so that the low end points north.* Turn the dial so that Box 0 points at the Sun. The Moon lies in the direction from the center of the dial to your mark.

Last-quarter Moon in Box 23

New Moon in Box 0

First-quarter Moon in Box 7

Angle equal to your latitude

Full Moon in Box 15

Mapping the Moon

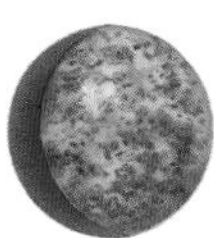

LOOK AT THE MOON with binoculars or a telescope, and you will see a fascinating mass of detail: mountains, craters, lava-flooded basins, and dazzling rays that cross miles of moonscape in straight lines. Early astronomers thought that the dark patches they saw were seas, which they gave such names as the Ocean of Storms and the Bay of Rainbows. Although we know they are not seas, the names have stuck. Many of the thousands of craters were named after famous historical figures —which is why we find Julius Caesar, Pythagoras, and Copernicus on the Moon. The most accurate maps made of the Moon are assembled from images taken by orbiting space probes, pinpointing landing sites for the *Apollo* missions.

The history of lunar mapping

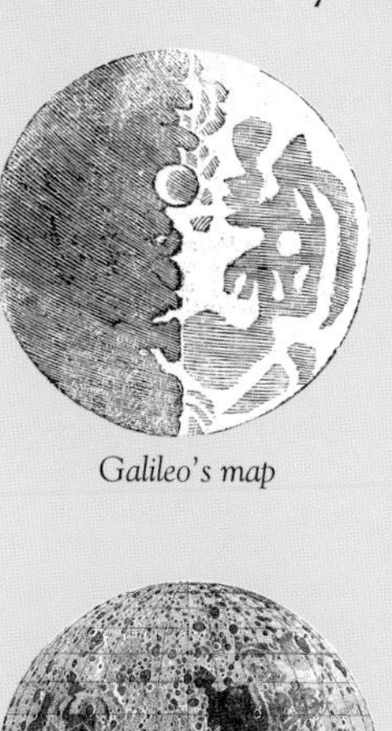

Galileo's map

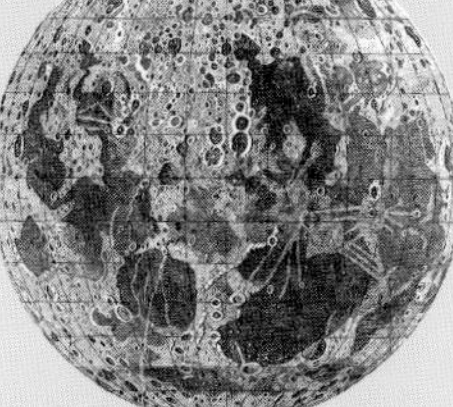

Tobias Mayer's map

Beer and Mädler's map

Galileo's first maps of the Moon were made in 1610 with his tiny telescope. These showed its battered, cratered surface for the first time. A Welsh astronomer of the time, Sir William Lower, described the Moon's surface through a telescope as looking like the "the crust of a tart." As telescopes improved in quality and increased in size, astronomers made more detailed maps of its "seas" and craters, such as that made in the mid-18th century by Tobias Mayer. One of the most accurate Moon maps was drawn up in the mid-19th century by Johann Heinrich Mädler and his colleague Wilhelm Beer. All these maps were made simply by carefully sketching the view seen through a telescope. In the 1960's NASA sent five *Lunar Orbiter* space probes to circle the Moon and map the surface in detail before the *Apollo* missions. Only some regions near the lunar poles have still to be mapped.

EXPERIMENT

Making a Moon map

You can observe the Moon with the naked eye, binoculars, or a telescope and then make your own map of the Moon. Check in your local newspaper to find when a new Moon is due, and observe it every night until it is a full Moon. After a full Moon, the Moon may rise too late for you to sketch, but you may catch it during the day.

YOU WILL NEED

- *binoculars or telescope*
- *pen*
- *scissors*
- *colored pencils*
- *glue*
- *paper*
- *poster board*

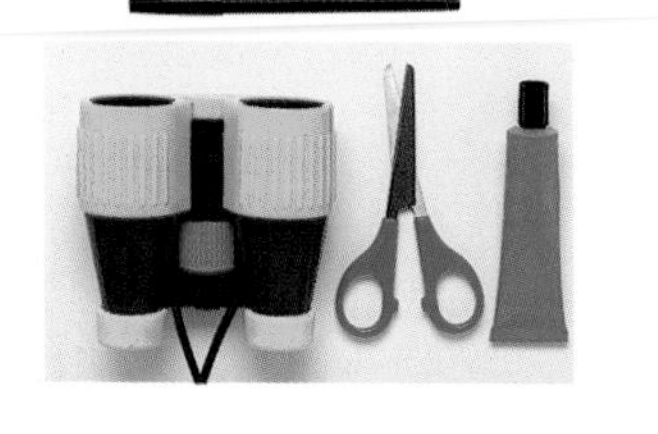

1 DRAW A LARGE CIRCLE 16 in (40 cm) wide on a piece of poster board. This is the beginning of your lunar map and will represent the Moon's surface, on which you can place the seas and craters that you see.

2 ON PAPER, COPY the lunar seas about twice as big as they appear opposite, and color them in. They are the easiest things to see on the Moon's surface, but they can look different as the Sun's light moves across them.

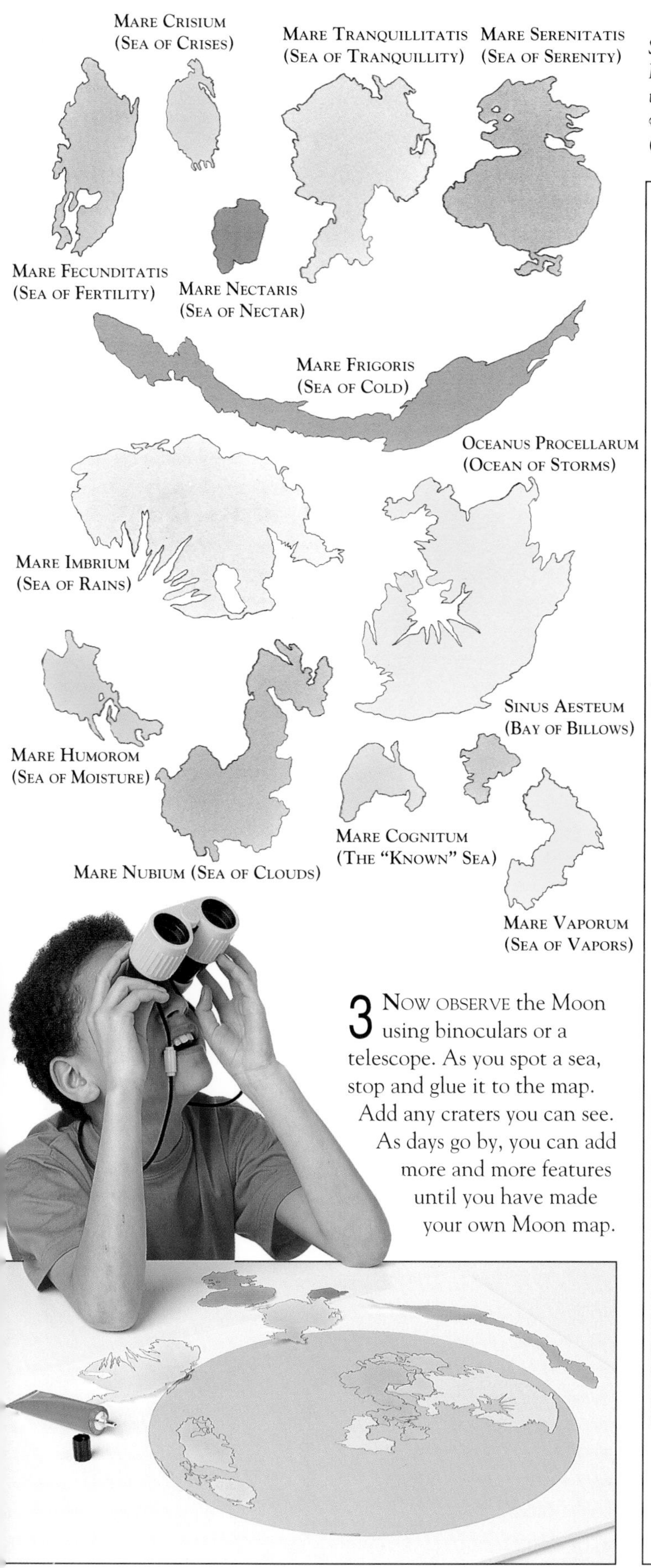

3 NOW OBSERVE the Moon using binoculars or a telescope. As you spot a sea, stop and glue it to the map. Add any craters you can see. As days go by, you can add more and more features until you have made your own Moon map.

Seas on the Moon
Lunar "seas" were formed by very large meteorites hitting with such force that they created gashes in the young Moon's surface. Molten lava then welled up and covered each gash's floor to a uniform depth. Copy each sea (mare in Latin) to use on your lunar map.

Lunar craters

Although the craters on the Moon were all formed by meteorite bombardment, they often look very different from one another. Some large craters have dark floors, because molten lava has flowed from inside the Moon to fill the deep cavities blasted out by meteorites. Very old craters may be covered by newer lava flows. Many craters have central mountains, caused by the ground "bouncing" back after impact. The youngest craters of all are surrounded by bright "rays" made of ejected material. These pictures show some common types of craters.

Gravity has made the walls of this crater slump

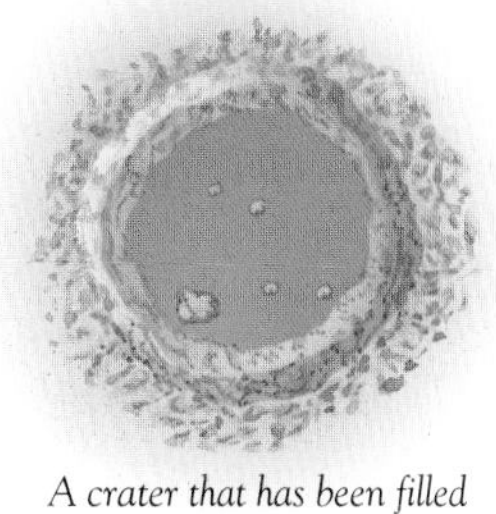

A crater that has been filled with lava welling up

A young crater with rays

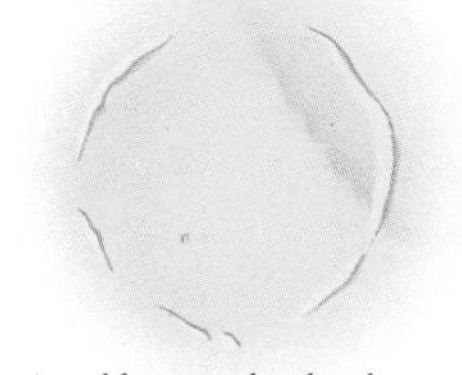

An old crater that has been submerged by lava

A small crater with a central peak

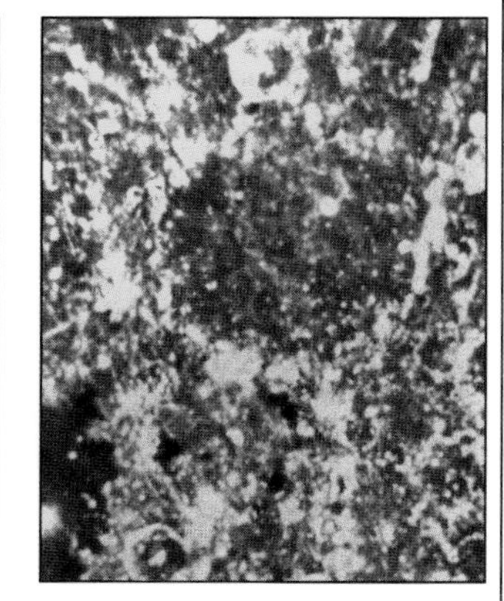

Now you see it, now you don't
Light hitting the Moon at an angle makes shadows that allow features to be seen. At full Moon the features cast no shadows as the Sun is directly above them. On the left are the craters Ptolemaeus and Alphonsus at last quarter; on the right are the same craters at full Moon.

The Moon's surface

THE MOON IS a very different world from the Earth. While the Earth is ever-changing, the Moon has not altered for billions of years—it still bears the crisply preserved scars of cosmic bombardment that ended 3.8 billion years ago. As the Solar System formed, debris from the creation of the planets hammered into all the young worlds. But while the Earth, Mars, and Venus were massive enough to retain their atmospheres and internal heat sources to renew their surfaces, the Moon was too small and simply fossilized after its period of cosmic shelling. The Moon's biggest craters measure hundreds of miles across, while many of the maria, or "seas" (p.49), are over 625 miles (1,000 km) wide. The maria were formed when giant meteorites hit the surface with such force that molten rock from inside the Moon welled up to fill the holes, then solidified.

EXPERIMENT

Measuring the heights of Moon mountains

You can easily see with a small telescope or a pair of binoculars that there are mountains on the Moon and high walls around the craters—but how high are they? Even before people went to the Moon, astronomers on Earth were able to work out the height of lunar mountains by measuring the length of the shadows they cast. The first person to do this was Galileo, in the early 17th century, and he obtained very accurate results. Astronomers know that the shadow of a lunar mountain is longest when the Sun is low on the Moon's horizon, and the shadow gets shorter as the Sun rises higher over the Moon's horizon—just as on the Earth. With the aid of mathematics, the exact height of the mountain can be found. You can try this on a smaller version of the Moon by using a "Moon mountain measurer" and completing the calculations.

YOU WILL NEED

- *thread* • *tape*
- *pen* • *glue*
- *piece of wood*
- *large ball*
- *poster board*
- *scissors*
- *tape measure*
- *notepad* • *flashlight*
- *ruler with millimeters*

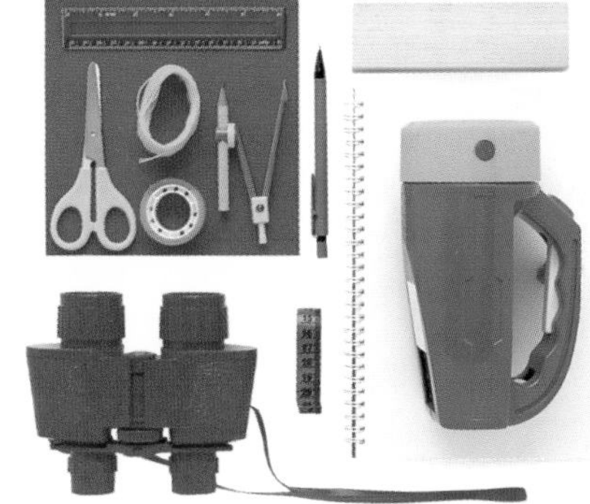

EXPERIMENT

Making craters

The Moon's craters were made by the impact of meteorites that hit the surface at high speed and then exploded. The energy that was released from these explosions created craters 20 times the size of the meteorites themselves. This is because both the meteorites and the surface became, in effect, liquid at the time of the explosion. You can make "craters" with plaster of paris.

YOU WILL NEED

- *plaster of paris* • *water*
- *bowl* • *spoon* • *tray*

Adult Supervision is advised for this experiment.

1 PLACE ABOUT 2 cups of plaster of paris in a mixing bowl. Gradually add water and stir until the mixture resembles a thick batter in consistency. Pour some onto the tray, and set the tray on some newspaper on the floor.

2 SCOOP UP A big spoonful of the remaining plaster mixture. Quickly, before it solidifies on the spoon, fling the mixture at the tray. How does the crater compare in size with the original spoonful of plaster? When you have finished, put the newspaper and plaster into the trash. Do not wash plaster down the drain, because it may clog the pipes.

1 MEASURE THE DIAMETER ("d") of the ball (the "Moon") by holding the tape measure over it. Cut out a small poster-board triangle (the "mountain"). Tape it to the ball.

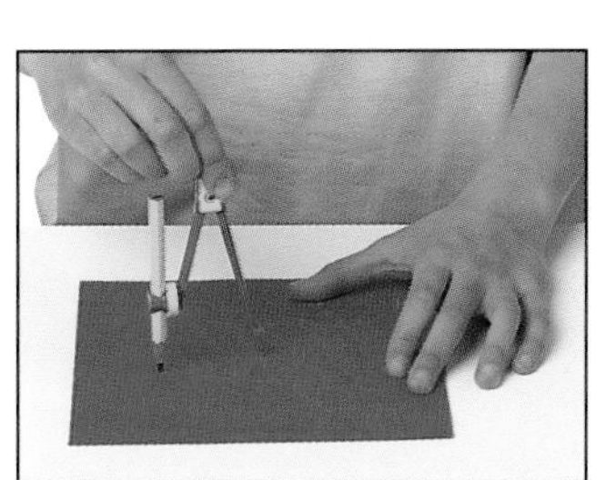

2 TO MAKE your Moon mountain measurer, first cut a 4-in (10-cm) diameter hole in poster board, and glue the poster board to the wood to make a stand. Adjust the measurer's position until the "Moon" fills the hole.

3 ASK YOUR FRIEND to shine a flashlight (the "Sun") at the ball so that the ball is lit like a first-quarter Moon. Turn out the lights in the room. The "mountain" will cast a shadow. Tape a thread to the measurer to mark the "Moon's" equator. Add a thread vertically, bisecting the hole—this is thread "L."

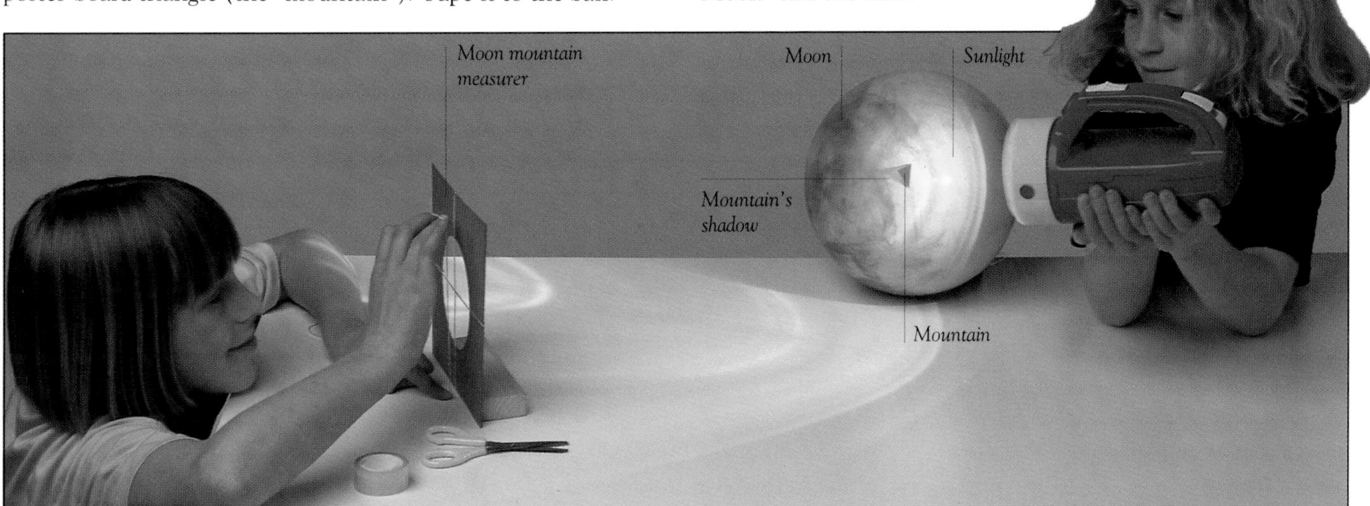

Making the calculation
Put two other threads vertically across the hole so that they cross the peak of the "mountain" and the tip of the shadow. Mark these threads "M" for mountain and "S" for shadow. Measure these distances in millimeters: the distance from thread M to thread L is "l"; from M to S is "s." Calculate the height of the "mountain" ("h") by using this formula: h = s x l x d / 8.

Where did the Moon come from?

No one is sure how the Moon was formed, but there are many theories about where it came from. Some of the most likely theories suggest that the Moon was formed from the Earth, that the Earth's gravity captured the Moon, or that the two were formed at the same time. The most popular theory today suggests that the Moon was the direct result of another planet crashing into the Earth. We must go back to the Moon and study it in detail to find the answer.

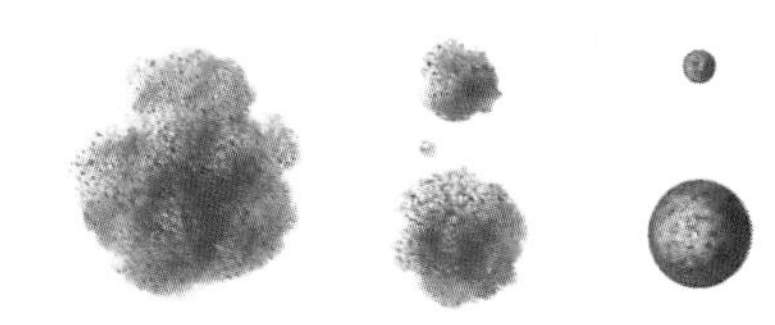

Solidifying clouds
Clouds of cosmic debris may have collected under gravity, forming two bodies that were close enough together to stay permanently paired.

The fission theory
The fission theory suggests that the Earth was spinning so rapidly that part of the planet began to bulge. This bulge separated to form both the Moon, which orbited the Earth, and Mars, which traveled out of the reach of the Earth's gravity.

Gravitational attraction
The Moon may have formed in another part of the Solar System. As the Moon traveled past the Earth, Earth's gravity attracted the Moon, which became trapped in orbit.

Cosmic collision
A body the size of Mars may have crashed into the Earth. The debris from the collision could have pulled together to form our Moon.

Gravity and the Moon

GRAVITY IS the force that makes a ball fall to the ground and keeps the planets in orbit around the Sun and the Moon around the Earth. An understanding of gravity is essential for tasks such as controlling a space probe and calculating how quickly the Universe will expand. The force of gravity between the Earth and Moon, however, has a far more immediate effect on us. The pull of the Earth on the Moon keeps the Moon in its orbit, and the pull of the Moon (and to some extent the Sun) on the Earth causes the tides. As the Earth spins "under" the Moon, the Moon tugs on our oceans, while the waters on the other side of the Earth swing out like a twirling dancer's skirt. As a result, two tides a day sweep around our planet. The Moon's gravity is also slowing the Earth's rotation down. The month and the day will eventually be the same length.

How the Moon causes the tides

The Moon's gravity pulls on all parts of the Earth, swinging the Earth around the center of gravity of the Earth-Moon system. The oceans on the side facing the Moon feel the Moon's pull slightly more strongly, so they are pulled up into a "bulge." The water on the other side feels the Moon's pull less strongly than the Earth as a whole, so the water there is flung outward into a second bulge. When the Sun and Moon are aligned, there are extra-high tides.

Low tide
The effect of the Moon's gravity can be very noticeable. At low tide the water level is at its minimum, because the Moon's gravity and the Earth's swinging motion have dragged the water to places that are thousands of miles away.

High tide
At high tide the Moon's pull seems to make the water rise up the shore. Tides come in cycles—every 2 weeks; when the Sun and Moon are aligned in our skies, they are at their highest ("spring" tides), with smaller "neap" tides in between.

EXPERIMENT

Measuring the acceleration caused by gravity

When Isaac Newton (p.38) realized that the Earth's gravity pulled on both the Moon and an apple, he wanted to know how strong that pull was. Other scientists had already measured how fast things fall, without understanding about gravity. They had found that falling objects go faster and faster (they accelerate) as they fall. This acceleration is the same for all objects, as long as the air resistance is the same. In this experiment you can examine how fast two objects of the same size and air resistance, but different weights, fall.

YOU WILL NEED
- *2 balls of equal size but different weights*
- *ruler*
- *watch*

Observing the effect of gravity
Ask a friend to hold the balls up as high as possible and to drop them at exactly the same time. Watch the balls drop. Do they accelerate at the same rate? Do they land at the same time? If you want to find out how quickly the balls accelerate, you can use the formula below.

Calculating the rate of acceleration
Measure the height from which the balls are dropped, and call this "s" feet. Time the balls as they fall three times, and take an average time—call this "t" seconds. Then use the following formula to find the rate of acceleration due to gravity ("g"): $g = 2s/t^2$. In other words, multiply "s" by 2, then divide the result by the amount you get when you multiply "t" by "t." This will give you "g" in feet per second per second for the Earth.

EXPERIMENT

What time is the tide due?

People who live near the sea or tidal rivers know that the tides do not come at the same times each day—in fact, the two tides in a day are separated by more than 12 hours. If the tide is high at 6:00 A.M., the next high tide is at 6:25 P.M., and the next morning high water comes at 6:50 A.M. The reason for this is that the tidal bulges are not fixed in position, but follow the Moon as it travels around the Earth once a month. So you have to move a bit farther around the Earth than you would expect before it catches up with the tidal bulge. Below, you can make a tide dial that will show you the progression of the tides as the Moon orbits the Earth and the Earth rotates on its axis.

YOU WILL NEED

- *poster board* ● *ruler*
- *compass* ● *paints*
- *paintbrush* ● *scissors*
- *paper fastener* ● *pen*

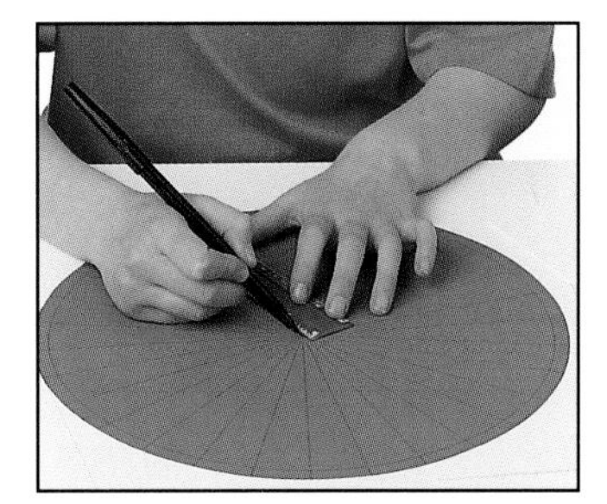

1 DRAW A CIRCLE with an 11-in (275-mm) diameter on poster board. Cut it out, and divide it into 30 equal segments. Label the segments from 1 to 30.

2 DRAW a $2\frac{1}{4}$-in (55-mm) circle, and paint it as the Earth. Draw a pointer $5\frac{1}{2}$ in (140 mm) long, and paint on tidal bulges and the Moon $2\frac{3}{8}$ in (60 mm) away, as below.

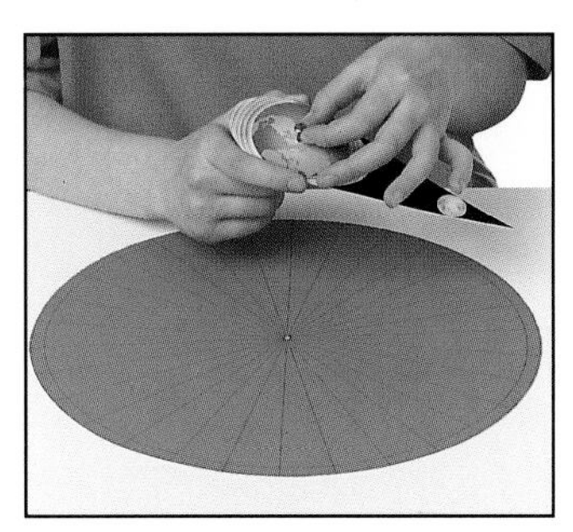

3 CUT OUT the pointer and the "Earth." Put "Earth" on top of the "tidal bulges" on the pointer, and push a paper fastener through both and through the center of the circle. Make sure that all the pieces can rotate freely.

4 ON THE DIAL LINE up the Moon so that the pointer points to "1." The fastener is the North Pole. Make a mark at one point on Earth's edge, and line it up with the tidal bulge facing the Moon. Move the Moon slowly counterclockwise, turning the Earth counterclockwise much more quickly, so that the Moon moves by one segment as the Earth turns once. The Earth must move by one extra segment before it reaches high tide again; this makes it about 50 minutes later than on the previous day.

Segments are labeled from 1 to 30 counterclockwise for the Northern Hemisphere

High tide

Low tide

The Moon revolves around the Earth, pulling the tidal bulges with it

The fastener represents the North or South Pole

Each segment represents 1 day of the Moon's orbit

The completed tide dial
This is what your tide dial will look like after you have assembled it. Imagine you are at the spot you have marked on the Earth. Where would you expect to see the Moon when the tide is high? When the tide is low?

Going to the Moon

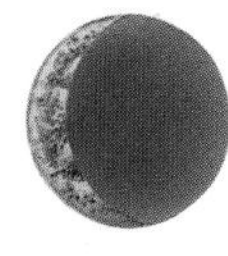

THE FIRST SPACECRAFT to reach the Moon was the Soviet *Luna 2*, which crash-landed on September 13, 1959. Since then, dozens of spacecraft have landed on, orbited, or traveled across the Moon, highlighted by the *Apollo* series of crewed lunar landings between 1969 and 1972. In those years 12 American astronauts lived, worked, drove, and even played golf on the Moon. They conducted experiments, brought back samples of moonrock for analysis, and learned how to live on another world—one with less gravity, no atmosphere, and vicious extremes of heat and cold. By early in the 21st century, there will probably be a permanent moonbase, where space travelers will experience living on another world for longer periods. The lessons we can learn on the Moon should help in the exploration of other worlds, such as Mars.

EXPERIMENT

The color of moonrocks

Moonrocks are a very dark brownish gray, and they reflect only 7 percent of the light that falls on them. A typical earthrock reflects 30 percent of the incoming light. The Moon looks bright in our sky only because we have nothing similar with which to compare it. If the Moon were made of earthrock, it would look far brighter.

YOU WILL NEED

- *pile of similar pebbles*
- *dark gray paint* • *paintbrush*
- *flashlight*

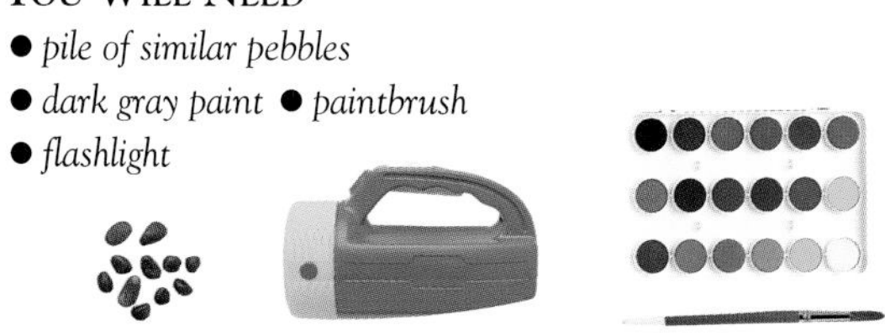

The *Apollo* missions

The first manned mission to land on the Moon was *Apollo 11* on July 20, 1969. Neil Armstrong and Buzz Aldrin walked on the Moon while Mike Collins remained in the orbiting capsule. The astronauts of the next six *Apollo* missions measured the Moon's magnetism and moonquakes, brought back one-third of a ton of moonrock, and drove a "lunar roving vehicle." Here John W. Young of *Apollo 16* is jumping in front of the United States flag.

EXPERIMENT

How much can you lift on the Moon?

Because the Moon is made of much less matter than the Earth, its gravity is much weaker. In fact, anything on the surface of the Moon weighs only one-sixth as much as it weighs on the Earth. This experiment will give you a sense of what it would be like to work in a lunar kitchen and the amounts you could lift if you were on the Moon.

YOU WILL NEED

- *empty containers* • *solid and liquid foods from your kitchen—be sure to ask permission before taking anything*

1 SELECT TWO EACH of a variety of unopened household foods. Empty the contents of one of each pair into its own separate container, then put a sixth back into the original packaging.

2 PICK UP THE unaltered food. Compare the sixth-full package to feel how light things are on the Moon. To find how much you could lift there, multiply the heaviest weight you can lift by six.

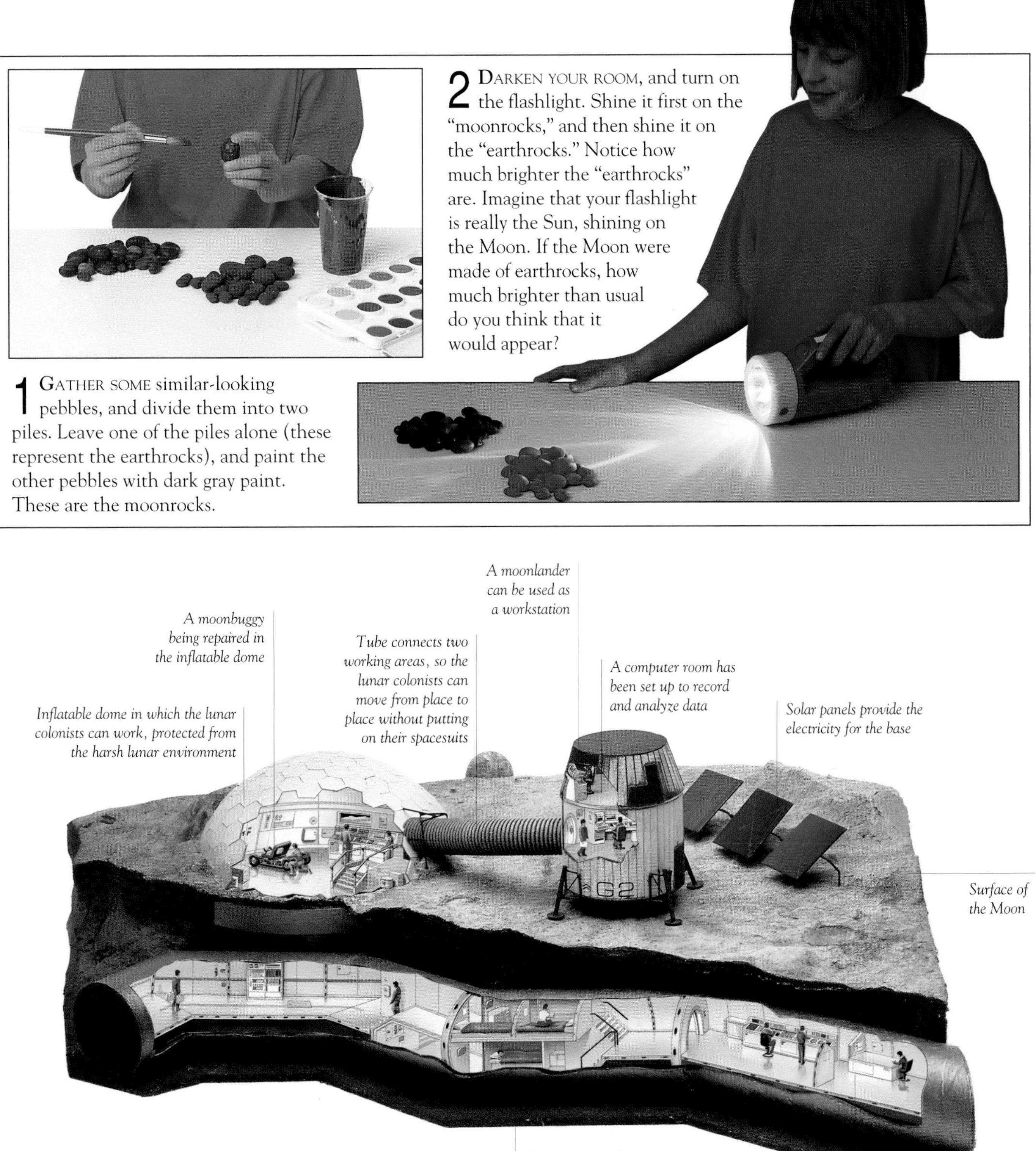

2 DARKEN YOUR ROOM, and turn on the flashlight. Shine it first on the "moonrocks," and then shine it on the "earthrocks." Notice how much brighter the "earthrocks" are. Imagine that your flashlight is really the Sun, shining on the Moon. If the Moon were made of earthrocks, how much brighter than usual do you think that it would appear?

1 GATHER SOME similar-looking pebbles, and divide them into two piles. Leave one of the piles alone (these represent the earthrocks), and paint the other pebbles with dark gray paint. These are the moonrocks.

A moonlander can be used as a workstation

A moonbuggy being repaired in the inflatable dome

Tube connects two working areas, so the lunar colonists can move from place to place without putting on their spacesuits

A computer room has been set up to record and analyze data

Inflatable dome in which the lunar colonists can work, protected from the harsh lunar environment

Solar panels provide the electricity for the base

Surface of the Moon

Living quarters for the colonists are protected by rock overhead

Working areas are built under the Moon's surface

Moonbase of the future

The 21st century may see the first permanent base on the Moon, for both scientific and commercial uses. Lunar rocks are rich in metals such as chromium, magnesium, manganese, and aluminum—all of which are useful for building new space stations and spacecraft. The low gravity of the Moon is perfect for making very pure drugs and electrical components that cannot be manufactured on the Earth. The dark, airless surface is also an ideal place from which to do astronomy, and the far side of the Moon has no radio interference from Earth. Most of the base will be underground to protect its inhabitants from extremes of temperature and the Sun's dangerous radiation. There, the lunar colonists will be able to work without spacesuits. They will also have the benefit of the Moon's low gravity and will weigh one-sixth of their Earth weight.

The SOLAR SYSTEM

Our backyard in space
Apart from Pluto, all the planets in the Solar System—and even a comet (entering from lower left and cutting through the orbits of the planets)—have been visited by spacecraft such as Voyager *(above). These missions have resulted in some astonishing discoveries about our neighbors in space.*

THE NINE PLANETS, 66 moons, and millions of asteroids and comets circling the Sun make up the Solar System—our local corner of the Cosmos. Space explorations like the *Voyager* probes have revealed some of the extraordinary variety among our neighbors in space—from the cloudy veil of Venus to the majestic rings of Saturn.

A PLANETARY FAMILY

THE SOLAR SYSTEM IS MADE UP of many different kinds of bodies: a typical star, the Sun; the planets Mercury, Venus, Earth, Mars, Jupiter, Saturn, Uranus, Neptune, Pluto, plus maybe more; more than 60 encircling moons; and literally millions of pieces of debris left over from the birth of the Sun and planets some 4.6 billion years ago.

Sky watching
This is a photograph of Venus (top) and Mercury (the tiny point of light at the bottom) with the crescent Moon in the early evening sky. They lie along a band across the sky called the zodiac. By looking in the newspaper at the beginning of every month you can find out which planets will be visible over the next few weeks.

The Sun is the most important body in our corner of the Universe, almost 1,000 times more massive than all the planets together. It is so massive that internal nuclear reactions make it shine (p.99). Planets, however, have no light of their own, and we see them because they reflect sunlight. Ever since ancient times, we have known of five planets whose gradual movement against the starry background revealed their relative closeness ("planet" actually comes from the Greek word for "wanderer"). The ancient Greek astronomer Ptolemy, whose teachings held sway for 1,500 years, thought the planets, Sun, and Moon circled the Earth, which lay at the center of the Universe. But in 1543 Nicolaus Copernicus, a Polish monk, proposed that the Earth and the planets move around the Sun. The Italian astronomer Galileo proved Copernicus's idea about 70 years later by observing the planets with the newly invented telescope.

Discovering planets
Clyde Tombaugh was the only person this century to discover a new planet. In 1930 he was working at the Lowell Observatory in Flagstaff, Arizona, when he found Pluto.

Three more planets, all very remote from the Sun, have been discovered since then: Uranus, Neptune, and Pluto. Until recently, the planets were only fuzzy disks seen through telescopes situated at the bottom of Earth's blurring atmosphere. Thanks to the space exploration programs, we have learned more about the planets in the last three decades than in all the time before. Since 1962 space probes have flown by (and in some cases, landed on) all the planets except Pluto—and now even a Pluto probe is under discussion. Just as navigators of the Renaissance period set out on voyages in search of new lands, so the space probes have revealed the planets as unique worlds.

Neptune's detection
In 1845 John Couch Adams predicted an eighth planet, beyond Uranus.

The inner planets

Mercury, closest to the Sun, is covered with craters caused by the bombardment of space debris. Venus has an atmosphere that is nearly 100 times heavier than our own, and it is made almost entirely of carbon dioxide, a gas that is extremely good at trapping the Sun's heat. As a result, Venus's surface temperature has risen to a searing 870° F (465° C), making it the hottest planet in the Solar System. Mars, the next planet beyond Earth, is in the grip of an ice age. Liquid water once flowed on the planet, and there may even have been oceans, but now the water is frozen into the soil. Mars has a thin atmosphere, polar caps made of ice, and dark markings that were once thought to be primitive vegetation. When the *Viking* space probes went to Mars in 1976, they found no life of any kind. A major problem is that Mars has no protective ozone layer, as the Earth does, to block the Sun's dangerous ultraviolet rays. Even if life started on the planet, it would probably not survive.

The outer planets

A zone of debris separates the inner planets from the next four planets, the gas giants. Having no solid surfaces, the gas giants cannot support life. In fact, life would be difficult anywhere near Jupiter, which has radiation belts

Venus's boiling atmosphere
The "greenhouse effect" on Venus is caused by gases trapping the Sun's heat energy in the atmosphere. Venus has a "runaway" greenhouse effect. In this experiment (p.67) you can create the same effect by sealing a jar and placing it in sunlight.

Rocks in space
This is Gaspra, a 12½-mile (20-km) long asteroid floating between Jupiter and Mars, photographed by the Galileo probe on its way to Jupiter.

(caused by electrically charged particles from the Sun becoming trapped in the planet's magnetic field) so strong that even space probes need to be protected. Jupiter's turbulent atmosphere heaves with violent storms, and Io, one of the planet's 16 moons, is riddled with active volcanoes.

All the gas giants are encircled by rings, but by far the most spectacular are the rings that surround Saturn. They are not solid, but are made of small chunks of ice, which orbit the planet like tiny moons. Saturn has moons too, and the biggest—Titan—has an atmosphere denser than the Earth's, with thick orange clouds. Beyond Saturn lie two worlds that at first appear to be twins —greenish Uranus and blue Neptune. Neptune is a warmer and much more active planet than Uranus. Floating among its blue clouds is the Great Dark Spot, a storm as big as the Earth. The biggest of its eight moons, Triton, is the coldest place in the Solar System—and yet it has active volcanoes.

Pluto's moon, Charon, is over half Pluto's size. Pluto and Charon may be the largest of a swarm of little bodies called "icy dwarfs" that are located in the far reaches of the Solar System.

The outer planets are made of the most common substances in the Universe, such as hydrogen, helium, and water. But nearer to the Sun, it is so hot that these gases and liquids have boiled away, so the inner planets are made mainly of rock.

Planetary debris

Pluto and Neptune do not mark the frontier of our Solar System. Some of the pieces of "debris" in our region of space—the comets—are thought to live in a giant spherical swarm beyond the planets, extending perhaps half the distance to the nearest star. These "cosmic icebergs" occasionally fall in toward the Sun, and they can put on a spectacular show as their outer layers evaporate. Still more debris is located in the asteroid belt between Mars and Jupiter. Sometimes fragments of asteroids stray from the belt, and some of these land on the Earth's surface as meteorites. By measuring the age of meteorites, we can date the formation of our Solar System to about 4.6 billion years ago.

Ancient meteorite
This is a slice through an iron meteorite that fell to Earth over 20,000 years ago.

Astronomers have evidence that the planets were a by-product of the birth of the Sun. A cloud of dust and gas in space shrank in size to form a spinning disk. The hot center became the Sun, while the dust particles in the cooler regions stuck together to make bodies the size of asteroids. These later came together to form the planets.

Astronomers are looking for planets that may be orbiting other stars, but the search is a difficult one because stars are extremely bright in comparison to the dim planets that orbit them. It is like looking for moths gathering around a distant streetlight. But astronomers have found that very young stars are often surrounded by hot, dusty disks that may be on the point of condensing into planets. So it is likely that a high proportion of stars like the Sun were born with a planetary family, and the number of planets in the Universe may be enormous.

Heavenly messengers
Comets have always fascinated people. Nearby comets, such as those shown in the old pictures here, have also caused terror. Some people believed that comets were the bearers of doom.

Jupiter explorer
The Galileo space probe is designed to explore Jupiter. While the main spacecraft orbits Jupiter, it drops a small probe into the seething atmosphere.

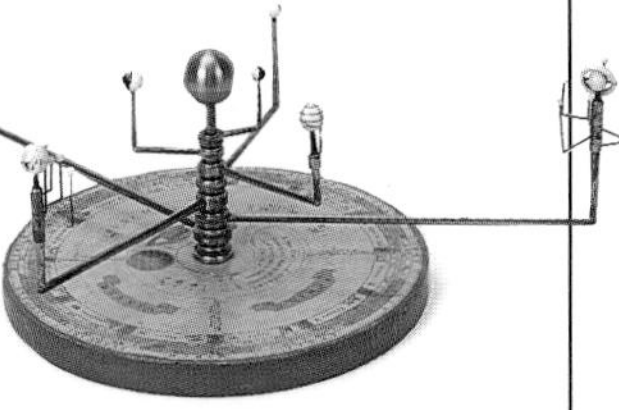

Watching planets
This strange-looking object is called an orrery. It shows how the planets in our Solar System revolve around the Sun.

The Solar System to scale

THE NINE PLANETS of the Sun's family come in a wide range of sizes, with the Earth near the middle. Our three nearest neighbors in space—Mercury, Venus, and Mars—are rocky worlds like the Earth, but they are smaller than our planet. The tiniest planet, Pluto, is a frozen ball of ice. The four "gas giants" (Jupiter, Saturn, Uranus, and Neptune) consist mainly of gases and water. These planets started out only a little larger than our own, but their gravity attracted extra gas from a huge cloud that surrounded the young Sun. This caused them to grow far bigger than the other planets, becoming giants in comparison.

EXPERIMENT

Solar System mobile

Making this mobile of the Solar System will show you the sizes of the planets to scale. The Sun, the center of the Solar System, is not a part of this model because of its enormous size—to be included, it would need to be 13 ft (4 m) in diameter. Also, remember that the real planets are not the same colors as in the mobile.

Diameters of the "planets"	
Mercury (pink)	*3/8 in (1 cm)*
Venus (orange)	*1 5/8 in (4 cm)*
Earth (light blue)	*1 5/8 in (4 cm)*
Mars (red)	*3/4 in (2 cm)*
Jupiter (yellow)	*19 in (48 cm)*
Saturn (crimson)	*15 in (38 cm)*
Uranus (dark blue)	*8 in (20 cm)*
Neptune (green)	*8 in (20 cm)*
Pluto (purple)	*1/4 in (8 mm)*

3 TAKE THE TWO circles for each "planet," and slot them together. Make sure the circles are at right angles.

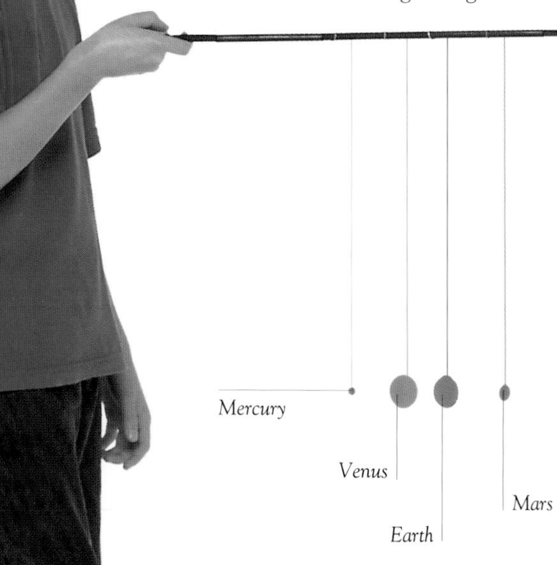

Our neighboring planets
Ask a friend to help you hold up the finished mobile. Make sure that the "planets" hang freely, and compare the amazing difference in their sizes. This mobile does not show to scale the actual distance of the planets from the Sun or from each other. To show this, the planets would be the same size, but the distances between them would be so large that you would need a pole several miles long.

EXPERIMENT

The sizes of the planets

From Earth, it is difficult to appreciate the three-dimensional sizes and volumes of the planets. Did you know it would take 1,300 Earths to make one Jupiter? In this experiment you can use foods to represent each planet. If you don't have the foods shown here, substitute other objects of equivalent sizes.

YOU WILL NEED

- *peppercorns (Mercury, Pluto, Mars)*
- *peas (Venus, Earth)*
- *plums (Uranus, Neptune)*
- *grapefruit (Jupiter)*
- *large orange (Saturn)*

Worlds in your hands
Put all the pieces of food on a table. Each item represents one of the nine planets of the Solar System. Hold them in your hands to compare their sizes. How many Plutos do you think it would take to fill Jupiter? How many Plutos would it take to fill the Earth?

YOU WILL NEED
- *poster board* • *ruler*
- *string* • *compass*
- *pencil* • *scissors*
- *pushpins*
- *long pole*

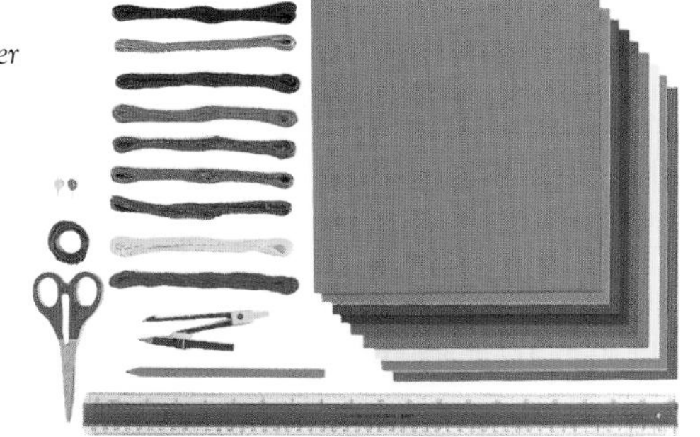

1 DRAW TWO circles for each "planet" (see chart opposite for diameters), using a compass or a string and pencil to make them the right size.

2 CUT OUT the circles you have just drawn. You should end up with 18 circles. Make a cut from the edge to the center point of each circle.

4 USE A PUSHPIN to make a small hole in the top of each planet, and thread some string through the holes.

5 ATTACH THE "planets" to the pole. Place them in the order below—the real order of the planets from the Sun. Leave enough space between them so that they do not bump into one another.

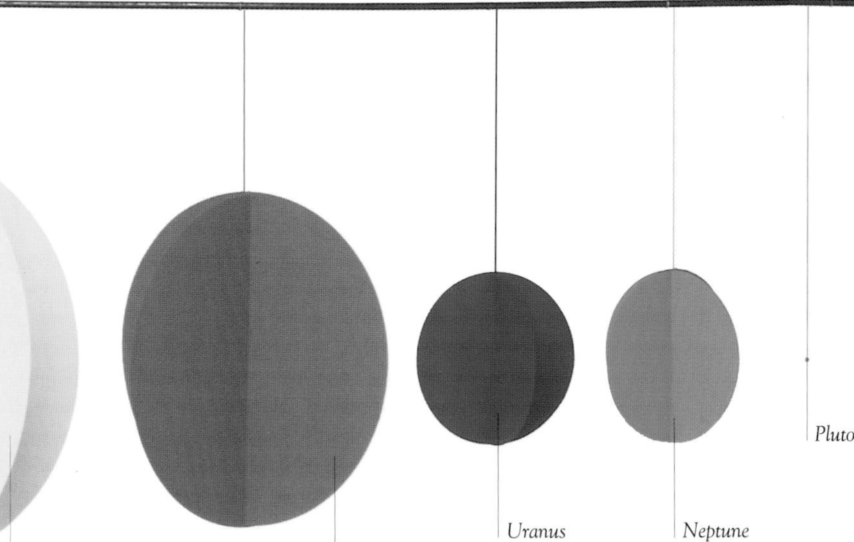

The planets: facts and figures	*Mercury*	*Venus*	*Earth*	*Mars*	*Jupiter*	*Saturn*	*Uranus*	*Neptune*	*Pluto*
Miles from Sun (millions)	36	67.2	93	141.6	483.6	886.7	1,783	2,794	3,666
Km from Sun (millions)	57.9	108.2	149.6	227.9	778	1,426	2,871	4,497	5,913
Mass (Earth=1)	0.055	0.81	1	0.11	318	95	15	17	0.002
Density (water=1)	5.4	5.2	5.5	3.9	1.3	0.7	1.3	1.6	2.0
Miles in diameter	3,031	7,520	7,926	4,217	88,846	74,898	31,763	30,775	1,419
Km in diameter	4,878	12,103	12,756	6,786	142,984	120,536	51,118	49,528	2,284
Gravity (Earth=1)	0.38	0.9	1	0.38	2.6	0.9	0.8	1.1	0.04

Planets on the move

LIKE THE EARTH, the planets orbit the Sun. In fact, the word "planet" comes from a Greek word meaning "wanderer," because all the planets are constantly on the move, traveling around the Sun along their ellipse-shaped orbits. The speed at which they travel depends on their distance from the Sun. Because the planets are always moving against the background of stars, their positions cannot be marked on ordinary star charts. To locate planets, check the monthly star maps in a newspaper or an astronomy annual.

Retrograde motion

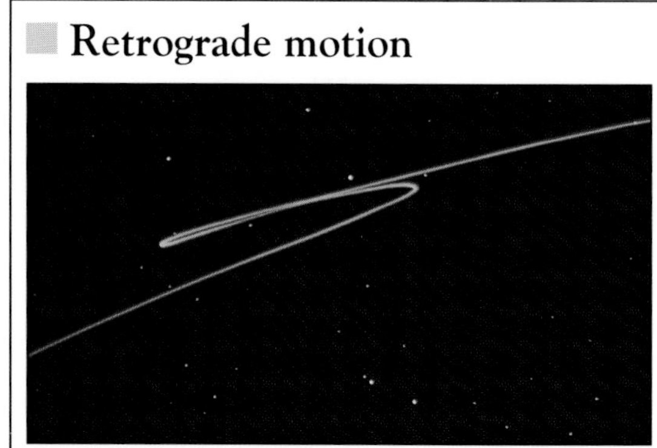

Sometimes, for a few weeks, planets beyond the Earth's orbit seem to drift backward in the sky. This is known as retrograde motion. The planet is still traveling forward but appears to fall behind as the Earth, in its faster orbit, overtakes it. This long-exposure photograph shows the retrograde "loop" made by Mars as seen from the Earth in the artificial sky of a planetarium.

EXPERIMENT

The motion of the planets

The planets orbit the Sun at different speeds depending on how close they are to the Sun. Mercury, the innermost planet of our Solar System, travels very fast to avoid being pulled in by the Sun's gravity. Pluto, the planet farthest from the Sun, travels much more slowly. Pluto's year—the time it takes to make one orbit of the Sun —is 248 Earth years long.

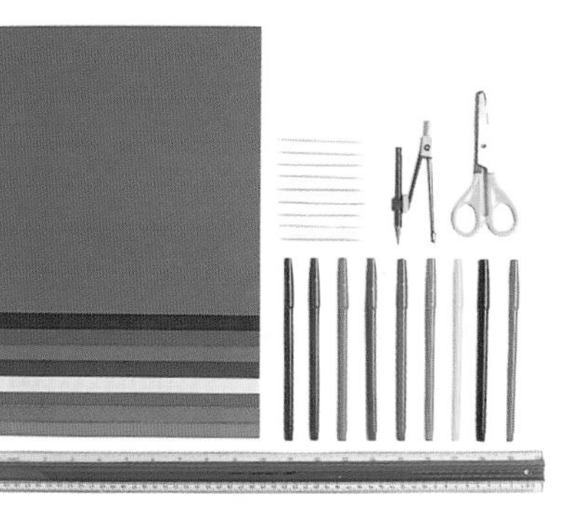

YOU WILL NEED

- *foamcore* • *compass*
- *colored pens*
- *colored paper*
- *toothpicks*
- *scissors*
- *tape*

Time taken by the planets for one orbit

Mercury	*3 months*
Venus	*7 months*
Earth	*12 months*
Mars	*23 months*
Jupiter	*142 months*
Saturn	*354 months*
Uranus	*1,008 months*
Neptune	*1,978 months*
Pluto	*2,976 months*

1 USE A COMPASS (or a string and pencil) to draw a Sun and nine circles around it, like a target, on the foamcore. Draw each circle in a different color.

2 MARK DOTS for the months (given above) in each planet's orbit. For the outer planets (beyond Earth), make just 12 closely spaced marks for each.

3 MAKE A COLORED FLAG to match each circle by cutting out a triangle from colored paper and taping it to a toothpick.

■ DISCOVERY ■

Johannes Kepler

Johannes Kepler (1571–1630) was a German mathematician who worked out three fundamental laws of planetary motion: (1) planets orbit the Sun in ellipses, (2) each planet moves faster when it is closer to the Sun, and (3) the length of each planet's year depends on its distance from the Sun. Half a century later, Isaac Newton proved that the underlying cause is the force of gravity.

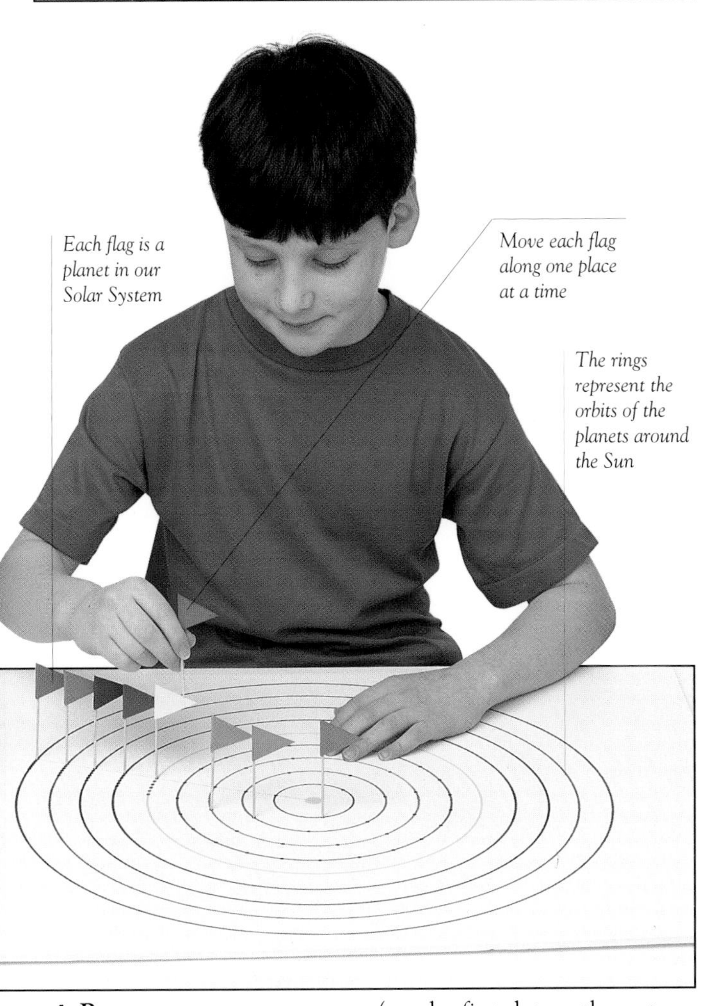

4 PLACE EACH FLAG ON A DOT (on the first dot on the outer circles). Move each flag counterclockwise by one dot, then another, until they have all traveled 12 dots. Which flags made more than one orbit? Which went around exactly once?

EXPERIMENT

How the outer planets move

Viewed from the Earth, the motion of those planets farther from the Sun than we are can seem confusing because they sometimes seem to change direction, an effect astronomers call retrograde motion (see opposite). This experiment, which you need to do with a (preferably taller) friend, shows how an outer planet can seem to move backward. You will be the Earth, and your friend the outer planet. The pole represents the Sun, around which all the planets move. Tie your rope loosely to the pole, so that it will not wrap around it, and hold on to the rope's middle. Your friend should tie his or her rope a bit higher and take hold of its end. Now, both move around the pole in the same direction, with you running and your friend moving more slowly.

YOU WILL NEED
- *2 equal lengths of rope*
- *a post or pole*

Planetary motion
Notice that you, the "Earth," circle the "Sun" much more quickly than your friend, the "outer planet." Outer planets, in larger orbits, have farther to travel, but they feel the Sun's gravity less strongly and do not have to move so fast. Notice the view from "Earth" as you overtake your friend: Does he or she seem to be moving forward or backward?

Mercury

THE CLOSEST PLANET to the Sun, Mercury speeds around its oval-shaped orbit in just 88 Earth days—less than a quarter of the time it takes the Earth to complete an orbit around the Sun. But despite its quick orbit, Mercury rotates on its axis so slowly that the Mercurian day is longer than its year. Mercury is the second smallest planet in the Solar System (only the distant Pluto is tinier), and it has one-third the diameter of Earth. It has a large core made of iron, which generates a magnetic field. The uncrewed spacecraft *Mariner 10* visited Mercury in 1974 and found that the planet is totally covered in craters and has virtually no atmosphere—just like our Moon. The surface of Mercury can become as hot as 840° F (450° C), but at night the temperature on the planet plunges to –275° F (–170° C).

EXPERIMENT

An elliptical orbit

All the planets move around the Sun in orbits resembling ovals, or ellipses. Most planets follow very rounded ellipses, but Mercury's orbit is quite elongated. Try drawing your own ellipse, as described here.

YOU WILL NEED
- *14-in (35-cm) long string*
- *foamcore* • *pushpins*
- *pencil* • *ruler*

1 DRAW A 5-in (12-cm) long line on the foamcore. At each end of the line, stick a pushpin into the foamcore. Tie the string into a loop.

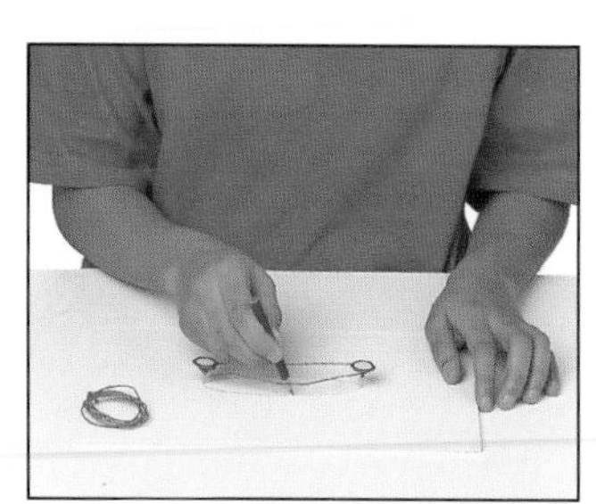

2 PUT THE LOOP of string around the pushpins. Place the pencil inside the loop, and pull it tight. Move it around the pins to draw an ellipse.

EXPERIMENT

The brightness of the Sun from Mercury

Because Mercury is so close to the Sun, the Sun looks very bright. The Sun appears nine times brighter from Mercury than from Earth—even though Mercury is only three times closer to the Sun, and the Sun looks only three times bigger from Mercury. This is an example of the Inverse Square Law (the brightness of a light source decreases with distance by an amount that depends on the square of its distance from you). Test this using two lamps. First, put them about 1 yd (1 m) and 3 yd (3 m) away from you. Cut a 1-in (3-cm) hole in two cards, cover with tracing paper, and place in front of the lamps.

YOU WILL NEED
- *2 lamps (with identical bulbs)*
- *tracing paper*
- *scissors*
- *ruler*
- *tape*
- *2 poster-board cards*
- *plastic putty*
- *2 wooden skewers*

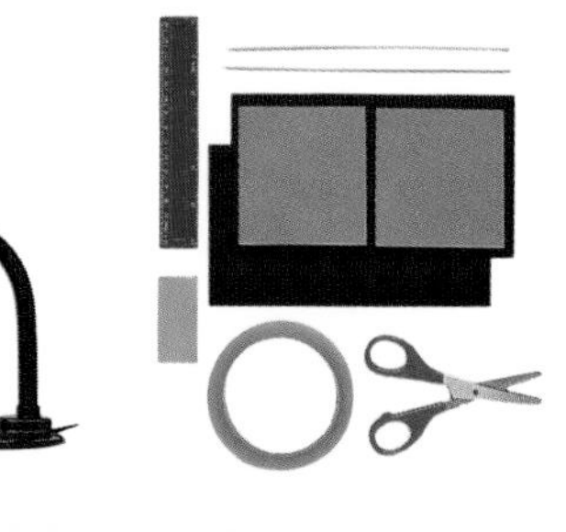

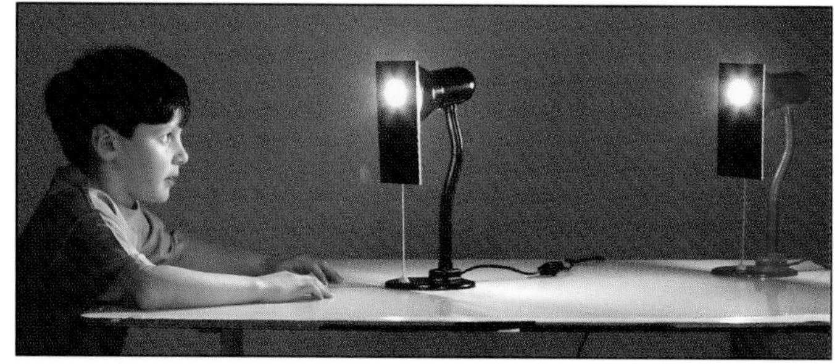

Comparing the brightness
Use skewers and putty to prop up the cards, and turn out the lights in the room. Then turn on the lamps. How different are they in size and brightness?

■ DISCOVERY ■

Einstein and relativity

During the 19th century, astronomers discovered that the oval orbit of Mercury is gradually swinging around in space. This perplexed astronomers until 1915, when Albert Einstein devised a new theory of gravity called the General Theory of Relativity. Einstein's theory predicted this motion for a planet so close to the Sun's strong gravity. Mercury helped prove that Einstein's theory was correct.

EXPERIMENT

Why Mercury's day is longer than its year

The period from noon to noon on Mercury (its day) is longer than the time the planet takes to orbit the Sun (its year). Here you can make your own model to show Mercury's strange rotation. The ball represents Mercury, and a modeling clay "snake" a mountain range.

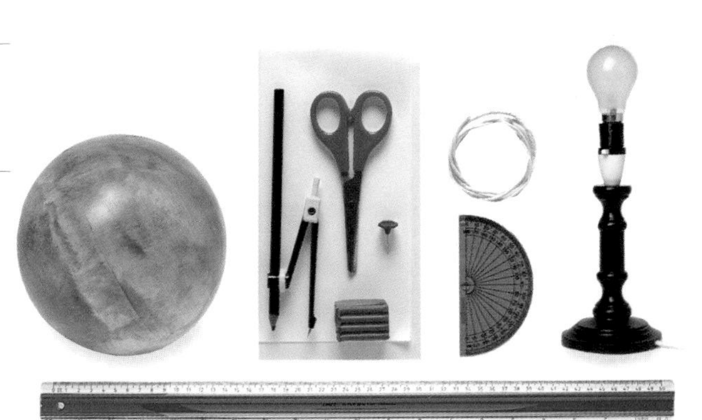

You Will Need

- *desk lamp with 40-W bulb* ● *paper* ● *poster board* ● *pushpin* ● *string* ● *pencil* ● *protractor* ● *ruler* ● *medium-size ball* ● *modeling clay* ● *scissors* ● *compass*

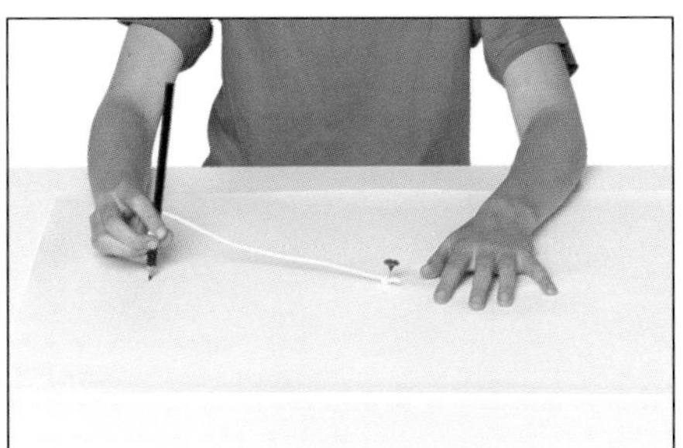

1 Tie a piece of string to a pencil, and stick a pushpin through the other end of the string and into a large sheet of paper. Now draw a circle. This roughly represents Mercury's elliptical orbit.

2 Now divide the circle into six pie-shaped segments. First, divide the circle in half, then divide the halves into thirds. Use a protractor to measure the inner angles—they should each be 60°.

3 Stick a "snake" of modeling clay to a side of the ball. Then, cut out 12 poster-board cards of a circle with a bump to match the outline of the ball and clay.

4 Place the lamp in the middle of the circle, and switch it on. Turn the lights out in your room. Place a card on the edge of the circle, on a line pointing toward the light. Place the ball on top, with the "snake" toward the light. Copy the shadowed "night" part of the ball on the card. Do this five more times, at each position moving counterclockwise, with the ball and a new card rotated counterclockwise by a quarter.

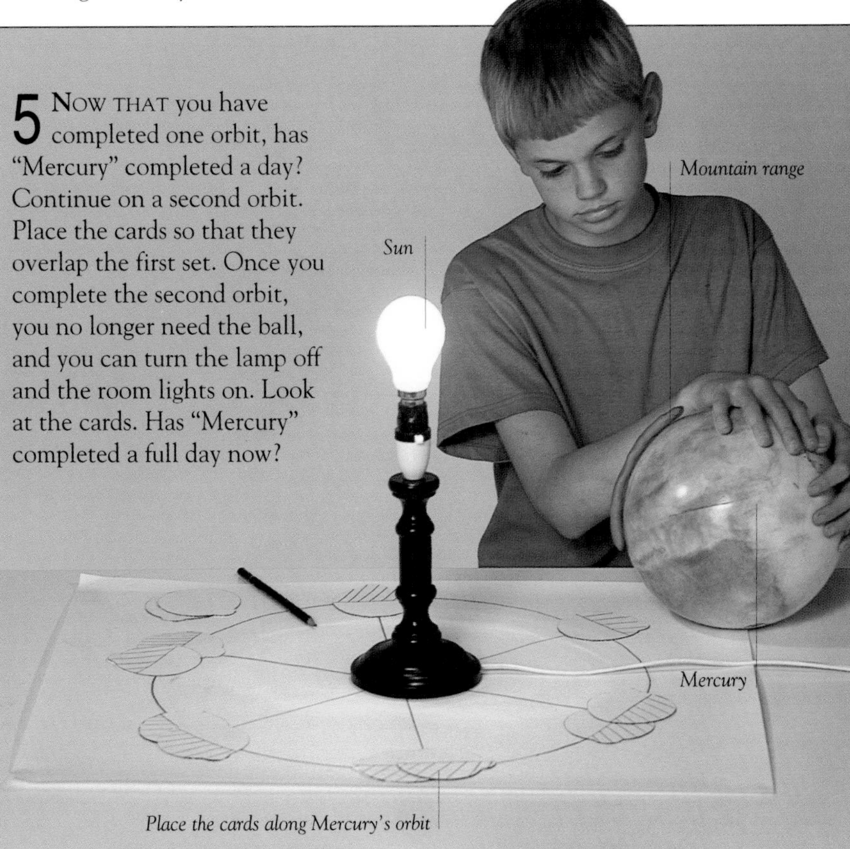

5 Now that you have completed one orbit, has "Mercury" completed a day? Continue on a second orbit. Place the cards so that they overlap the first set. Once you complete the second orbit, you no longer need the ball, and you can turn the lamp off and the room lights on. Look at the cards. Has "Mercury" completed a full day now?

Place the cards along Mercury's orbit

Venus

ALTHOUGH IT IS almost the Earth's twin in size, Venus is otherwise totally different from our friendly world. The "air" on Venus is made of unbreathable carbon dioxide, and it pushes down with a force that is 90 times stronger than the Earth's atmospheric pressure. This atmosphere traps the Sun's heat, raising the planet's temperature to 870° F (465° C)—so hot that the surface glows in the dark. Venus has huge volcanoes and lava lakes, which are hidden from the view of ordinary telescopes by a dense cover of acidic clouds. A visitor to Venus would be simultaneously suffocated, crushed, roasted, and corroded.

Seeing with radar
Astronauts cannot visit Venus in person, but radar waves can be sent to the planet. Then scientists can study the data and formulate an image of its surface. This computer image taken by the Magellan *spacecraft shows Maat Mons, a volcano on Venus that is 5 miles (8 km) high.*

EXPERIMENT

Radar mapping

Venus is completely covered with clouds, so the only way we can map its surface is by using radar waves. Radar penetrates the planet's atmosphere, hits the planet's terrain, and bounces back to a receiver. By studying the radar waves that have bounced back, scientists can determine the height and shape of the landscape beneath the clouds. In this experiment you can achieve the same result by making a simple "plumb line" to measure depth.

YOU WILL NEED
- *building blocks*
- *pen*
- *string* • *eraser*
- *large piece of poster board*

1 ASK A FRIEND to make a secret landscape of hills and valleys using building blocks. Do not look at what your friend has done.

2 TAKE THE PIECE of poster board, and draw a grid on it using the ruler and pen. Leave 2 in (50 mm) between each line. Ask your friend to put the grid in front of his or her landscape.

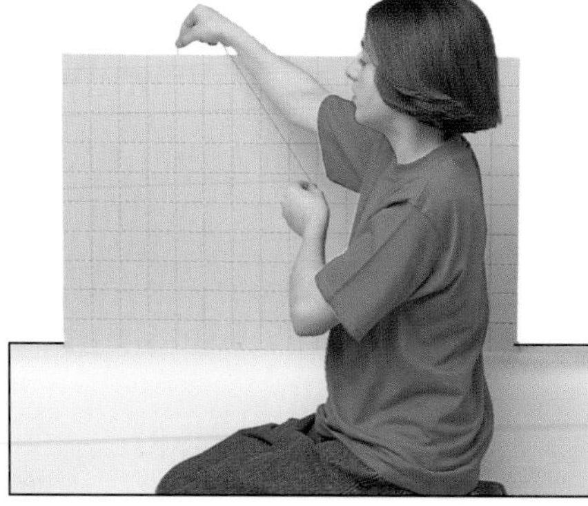

3 TIE A PIECE OF STRING around the eraser, leaving about 20 in (50 cm) of spare string. Without looking behind the grid, lower the eraser until it hits the landscape and cannot go any farther.

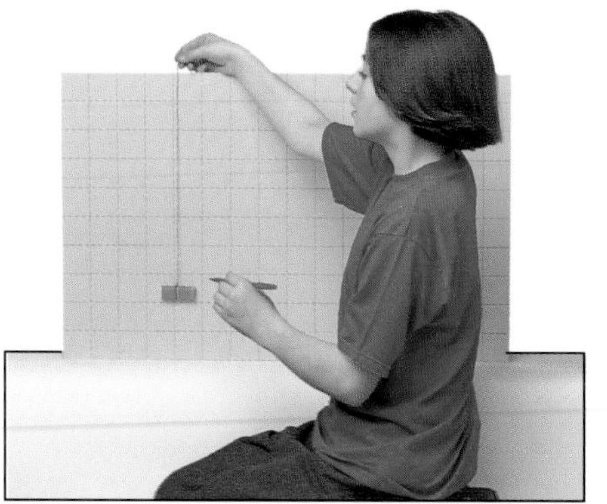

4 PINCH THE STRING at the top of the grid. Then bring the string to the same position as in Step 3, but in front of the grid. Use a pen to mark the position of the eraser.

Cloud blanket
A brilliant blanket of white clouds hides Venus's surface in this view from the Pioneer Venus Orbiter. *The clouds are made of sulfuric acid, erupted from the planet's active volcanoes, and reflect so much sunlight that Venus is one of the brightest objects in the sky. By studying Venus's atmosphere, we have learned much about our own.*

EXPERIMENT

The greenhouse effect

Mount two thermometers on poster board with plastic putty, and put them in sunlight, one inside a jar covered with plastic wrap. After an hour, compare the temperatures. The difference is due to the greenhouse effect: the Sun's heat is trapped in the jar. On Venus, the Sun's heat is trapped by gases in the atmosphere.

You Will Need

- *2 thermometers* ● *poster board*
- *plastic putty* ● *plastic wrap*
- *glass jar*

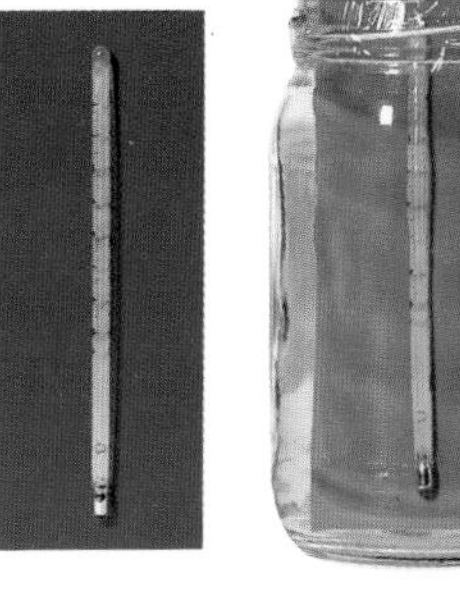

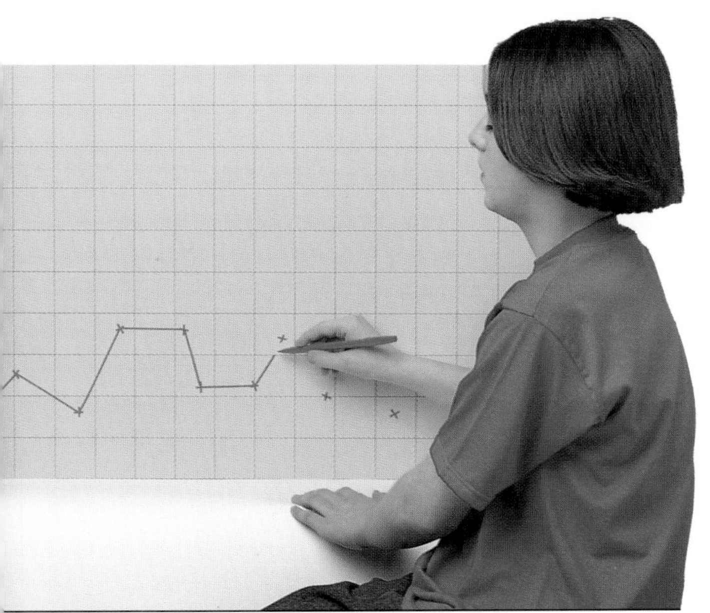

5 Repeat Steps 3 and 4, moving along the grid until you have marked at least 10 spots. Then use your pen to connect them. Without seeing the landscape, you have created your own "radar" map. Now, move the grid and compare your map with the landscape behind it.

EXPERIMENT

Phases of Venus

As seen from the Earth, Venus seems to change its shape, much as the Moon does. This was noticed by Galileo, the first person to point a telescope at Venus. You can see the phases of Venus by using binoculars or a small telescope, or you can follow this experiment to demonstrate the same effect.

Adult supervision is advised for this experiment.

You Will Need

- *foamcore* ● *compass*
- *scissors* ● *small light bulb*
- *battery* ● *2 wires with clips*
- *ball* ● *toothpick*
- *screwdriver* ● *pencil*

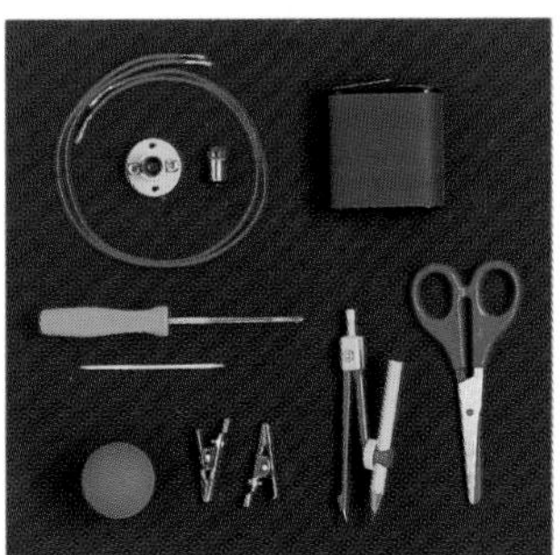

1 Use a compass and pencil to draw a circle with a 6-in (150-mm) diameter on a piece of foamcore. Then carefully cut out the circle using a pair of scissors.

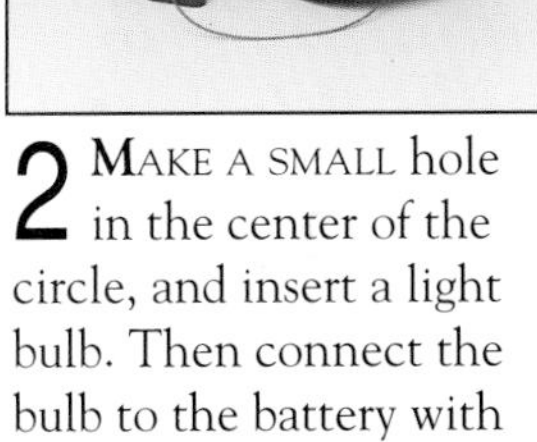

2 Make a small hole in the center of the circle, and insert a light bulb. Then connect the bulb to the battery with clips and wires.

3 Make a small hole in the ball, and insert the toothpick. Put the end of the stick into the outer edge of the foamcore, so that the ball is held upright.

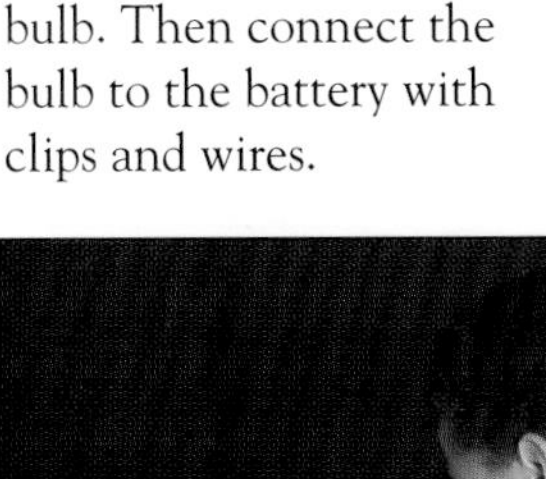

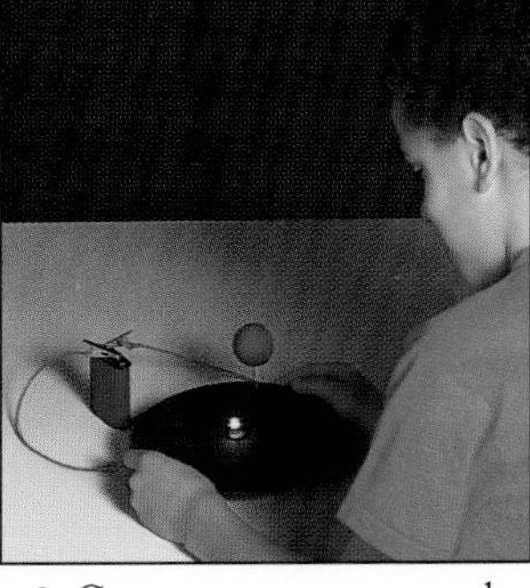

4 Connect the clips to the battery to light the bulb. Then darken the room, and gradually revolve the ball ("Venus") around the light ("Sun") to see the phases.

Mars

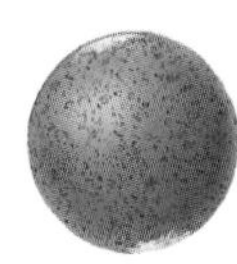

ALTHOUGH IT IS only half the size of the Earth, Mars—the "Red Planet"—has a volcano three times the height of Mount Everest and a huge canyon that could swallow up the Alps. Its two moons are probably captured asteroids (p.152). Since Mars is farther away from the Sun than the Earth is, it is cold with temperatures rarely rising above freezing. In the 19th century some astronomers thought they had detected a network of canals on the planet (p.71). These "canals" led some people to believe that intelligent life might exist on Mars, and there were many scary stories about Martians coming to invade the Earth. But although water—even oceans—may have covered Mars in the past, today there is only frozen red soil.

Olympus Mons
Towering above the thin Martian clouds, Olympus Mons is far bigger than any volcano on Earth—although it is no longer active. The base of the volcano covers an area larger than Spain. This view is from the Viking *spacecraft.*

EXPERIMENT

Build your own shield volcano

Mars is home to the biggest volcano in the Solar System, Olympus Mons. The volcano grew to an enormous size—16 miles (26 km) high and 375 miles (600 km) wide—because the crust of Mars does not move. Earth's crust is broken into fragments that "float" on the molten rock underneath. The floating fragments—"tectonic plates"—move slowly. Lava from a hot spot deep in the Earth can punch through the plate to make a volcano, but the moving plate takes the volcano away before it grows as big as Olympus Mons. Here you can compare Olympus Mons to the Earth's biggest volcano, Mauna Kea.

YOU WILL NEED
- *blue and orange poster board*
- *pen* ● *compass* ● *paper* ● *scissors*
- *ruler* ● *plastic putty* ● *plaster of paris with blue and orange powder paint* ● *spray bottle* ● *water* ● *scale*
- *spoon* ● *blue paint and brush*

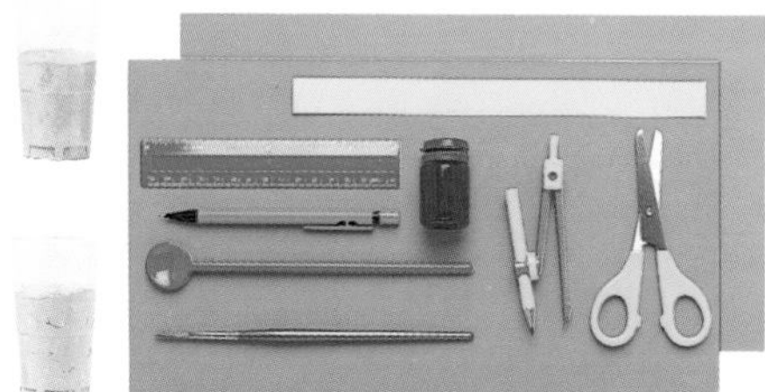

1 WEIGH EACH piece of poster board. On the orange piece ("Mars"), draw a circle with a 6-in (15-cm) radius. On the blue piece ("Earth"), draw a circle with a 2-in (5 cm) radius.

2 FOR "MARS," cut a strip of paper ½ in (12 mm) long; for "Earth," ¼ in (5 mm). These show the height of each model. Stand the strips in a blob of putty in the center of their respective circles.

3 MATCH THE colored plaster of paris mixtures to the bases, and make mounds ("volcanoes") to the heights of the strips and the widths of the circles.

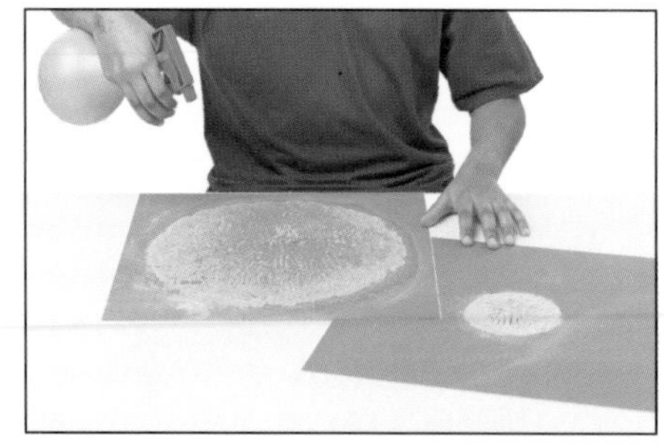

4 USE THE SPRAY BOTTLE to dampen the outside of each "volcano," and leave both to dry. Paint a blue border around "Mauna Kea."

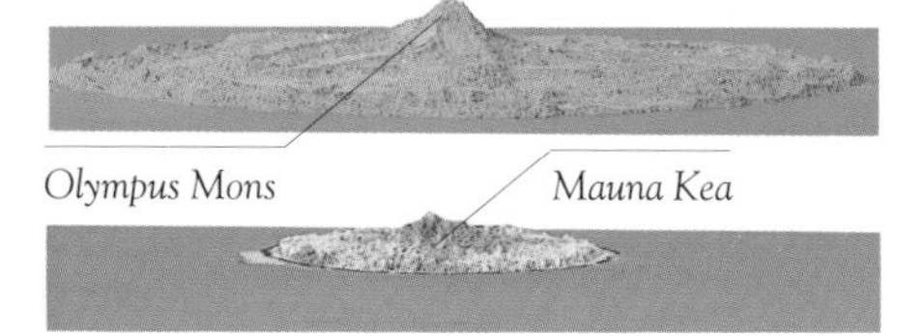

5 HOW MUCH MORE massive is "Olympus Mons" than "Mauna Kea"? Weigh each mound on its base. Subtract the original poster-board weight from the final weight to find the weight of each "volcano."

EXPERIMENT

Why Mars is red

Iron compounds in the Martian soil have rusted, turning Mars into the Red Planet. Water rusts iron, as this experiment shows. The color of Mars and the dried-up streambeds on its surface indicate that there was once water on the planet.

You Will Need

- *baking dish* • *sand* • *steel wool* • *scissors*
- *dishwashing gloves* • *pitcher of water*

Adult supervision is advised for this experiment.

1 HALF-FILL a baking dish with sand. With gloves, snip the steel wool into 1-in (2-cm) pieces, and mix it in.

2 POUR ENOUGH water into the dish to just cover the sand and steel-wool mixture. Leave the dish in a safe place.

3 CHECK ON THE DISH EVERY DAY. As the water evaporates, add a bit more to keep the mixture moist. Check the color after 3 days. How long do you have to leave the mixture until it is the same color as Mars?

EXPERIMENT

The canals on Mars

In the 19th century, astronomers using old-fashioned telescopes noticed long, straight features crisscrossing Mars. Some people thought these markings were canals, built to irrigate the planet. The "canals" turned out to be optical illusions caused by the eye "connecting" the dots.

You Will Need

- *red paper* • *black paint* • *paintbrush* • *pen*
- *binoculars* • *scissors* • *compass* • *tape*

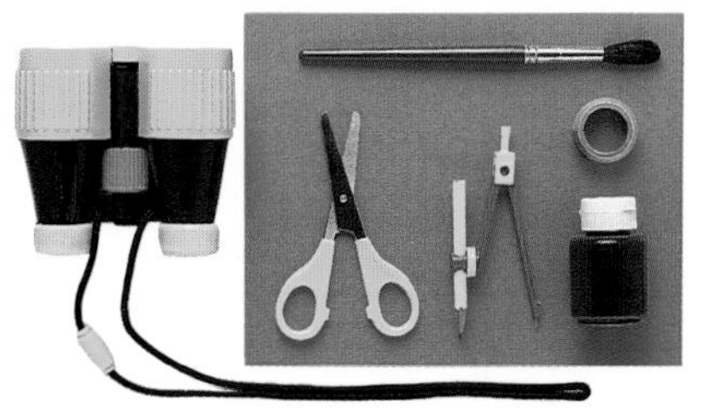

1 USE THE PAINTBRUSH to flick some black paint onto the paper. When the paint has dried, use a compass to draw a 6-in (15-cm) wide circle, then cut out the circle. This represents Mars.

2 TAPE "MARS" to a wall about 10 ft (3 m) away. View it through out-of-focus binoculars (this simulates looking at Mars through the blur of our atmosphere). What do you see?

Exploring Mars

Of all the planets in our Solar System, Mars is the most like Earth. Although it is a much smaller and colder planet, it has a thin atmosphere to protect its surface from space. It also has polar caps made of ice and seasons like our own. But does it have life? Even after people stopped believing in humanlike "Martians," some astronomers still thought that Mars was home to simple plants. Early space probes in orbit around the planet discovered dried-up river valleys—clear evidence that water, essential for life, had once flowed on Mars. And in 1976, NASA sent the *Viking* landers to find out more about the "Red Planet."

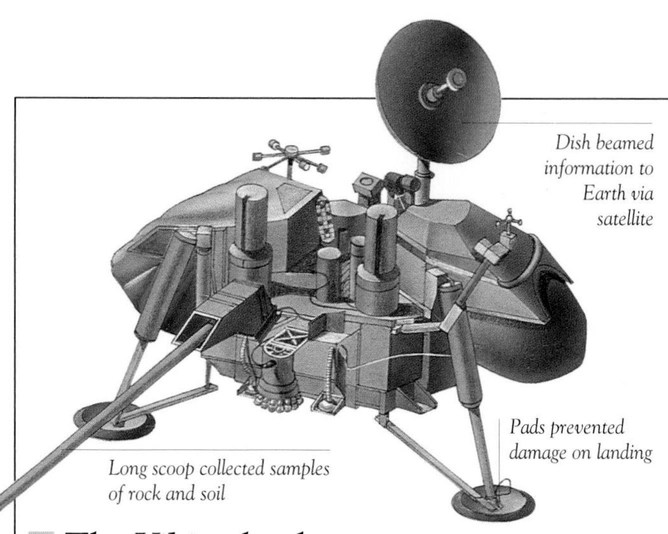

The *Viking* landers

No human has ever been sent to Mars to explore the planet, but two sophisticated space probes were sent there to collect information. They were the *Viking* landers, and they were able to "touch," "smell," "see," and "taste" the soil and monitor the weather. Scientists had hoped that microscopic life might be present on Mars, but they found only a chemically reactive soil.

EXPERIMENT

Testing for life

This experiment mimics tests the *Viking* landers made on Martian soil. The soil is fed with nutrients and observed for signs of life. Any chemicals in the soil will react quickly, but briefly. If living cells are present, they will react slowly and will continue to react as they multiply.

You Will Need

- *3 glass jars* • *clean sand*
- *long-handled teaspoon* • *2 tsp salt*
- *2 tsp yeast* • *2 tsp baking powder*
- *3 colored labels* • *large glass pitcher*
- *½ cup sugar* • *2 cups warm water*

1 Fill the jars one-third full of sand. Attach a different label to each. Mix the salt in the red jar, the baking powder in the yellow, and the yeast in the blue.

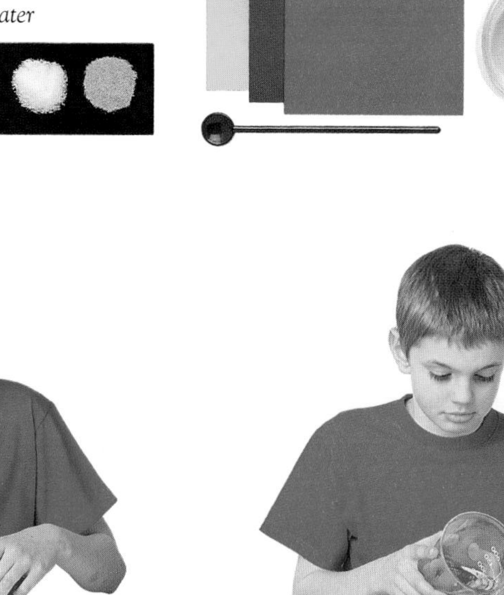

2 Place the jars in the refrigerator overnight, to imitate the cold on Mars. In the glass pitcher, mix the sugar and warm water well.

3 The next day, pour equal amounts of sugar water—the nutrient for the soil—into each jar. Set the jars aside, and watch for any reactions.

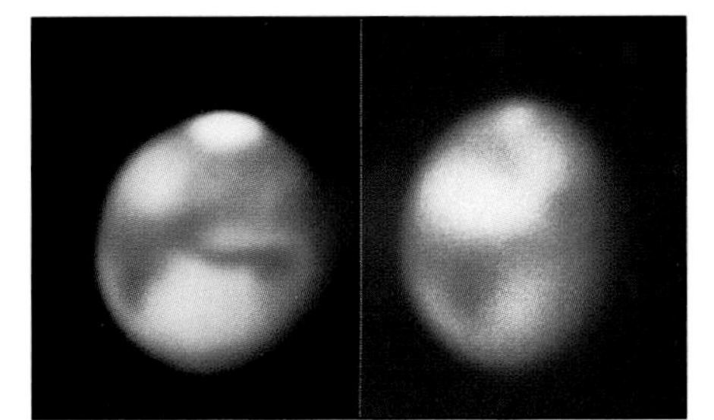

The changing face of Mars
Both of these pictures show Mars, though they look different. The right-hand image shows how Mars usually looks from Earth. The left-hand image shows a huge dust storm covering much of the bottom of the planet. This particular storm occurred in 1971 and lasted 5 months. The changing appearance of Mars was a puzzle to astronomers for years. Some people thought it was due to the growth and retreat of vegetation with the seasons. But with advances in telescopes and information from the Viking landers, we now know that the differing views of Mars are caused by the planet's turbulent weather.

EXPERIMENT

The wave of darkening

When dust blows back and forth across the Martian desert, rocks are covered and uncovered, changing the way the planet looks. To copy this effect, use a spoon to mix equal amounts of plaster of paris and powdered cocoa with enough water to form a paste. Spread the paste unevenly on a tray to represent the surface of Mars, and let this dry. Then, evenly sprinkle a handful of sugar over the tray. Blow over the tray from one end to reveal the dark surface below. Now blow from the other end. This is what happens halfway through the Martian year when the winds change direction.

YOU WILL NEED
- *large tray*
- *plaster of paris*
- *powdered cocoa*
- *water*
- *sugar*
- *teaspoon*

4 WITHIN AN HOUR you will have the results of your test. The contents of one jar will not have reacted with the nutrient at all. However, the contents of both of the other jars will have reacted, but in different ways. Which jar do you think contains life?

Jupiter

JUPITER IS the largest planet in the Solar System—it is so big that it could contain all the other planets put together. Because of its size, it is easily visible from Earth. Jupiter is made mainly of a liquid mixture of hydrogen and helium, and the planet has no solid surface. Its cloud patterns include the Great Red Spot, a storm that is three times the size of the Earth. The 4 largest of Jupiter's 16 moons are each bigger than the planet Pluto.

Observing Jupiter from the Earth

Jupiter and its biggest moons make an interesting sight through a small telescope or even through a pair of binoculars. This image of the planet was photographed through a small telescope. To observe the planet, support a telescope or binoculars firmly by resting your elbows on a flat surface, then focus on Jupiter. To either side of the planet, you should spot some tiny points of light. These are Jupiter's four biggest moons, and they change their positions from night to night. You may also be able to see Jupiter's bands of clouds.

EXPERIMENT

Jupiter's bulging waistline

Jupiter is noticeably wider at its equator—so much so that the planet's "bulging waistline" is visible even through binoculars. This extra width is caused by centrifugal force, which is the result of Jupiter's incredibly rapid rotation. Jupiter has the fastest spin of any planet in the Solar System—a day on Jupiter lasts less than 10 Earth hours. This experiment demonstrates how centrifugal force affects Jupiter.

YOU WILL NEED

- *wooden skewer* • *rubber band* • *plastic putty*
- *drinking straw* • *colored paper* • *tape*
- *scissors* • *ruler* • *pen or pencil*

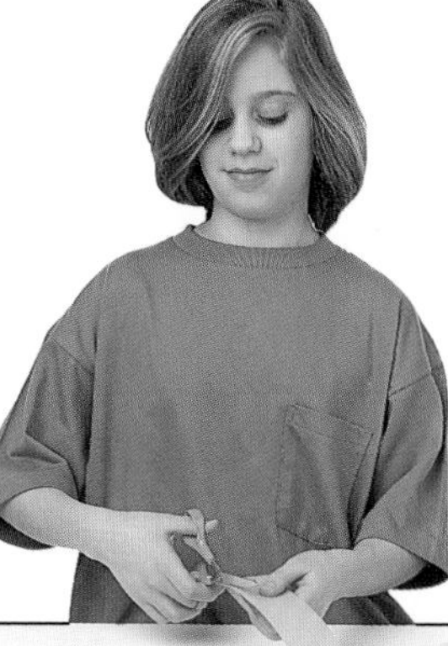

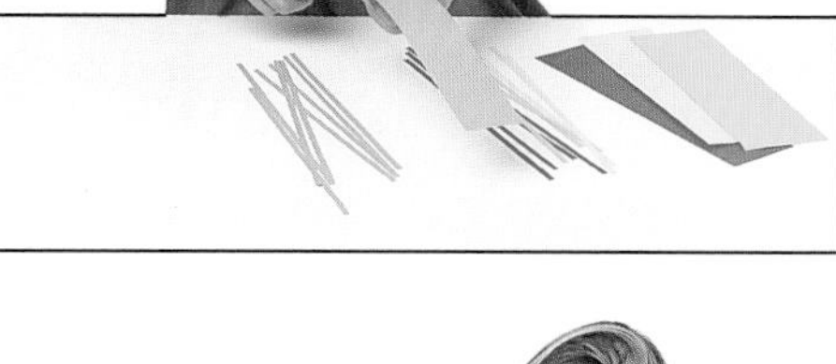

1 CAREFULLY cut out 20 to 30 strips of colored paper, each measuring 1/4 x 12 in (6 x 300 mm). Next, cut two small sections of the drinking straw, each about 1 in (25 mm) long.

2 AT EACH END of the skewer, slide on a piece of the straw. Then use some tape to attach the ends of the strips of colored paper to each of the straw pieces.

3 ONCE YOU HAVE fastened the strips of paper at both ends, cover the tape with a strip of paper. Make sure the straws move freely. Put a large blob of plastic putty on one end of the skewer.

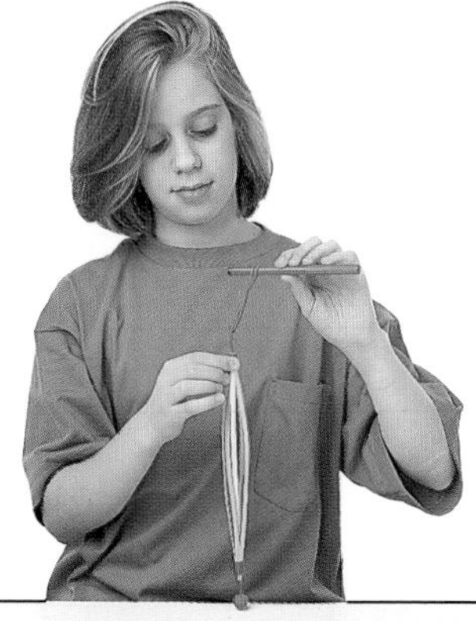

4 NOW TIE A rubber band to the opposite end of the skewer with some plastic putty to hold it, and tie the free end of the rubber band to a pen or pencil. This is the handle.

EXPERIMENT

Jupiter's magnetic field

All the planets are immersed in the Sun's huge magnetic field. Jupiter also has a strong magnetic field, which forms a large "bubble" around it. This bubble is invisible, but spacecraft have measured its extent and force. In this experiment you will draw a bubble around a magnetized nail.

You Will Need
- *cardboard box*
- *nail or screw*
- *magnet* ● *pencil*
- *pen* ● *compass*
- *plastic putty*

1 Make a hole in one side of a box, and push a pencil through. Stroke the nail with a magnet, and attach the nail to the top of the pencil with putty.

2 Move a compass from the corner of the box toward the nail until the needle shakes. Mark this spot. Repeat, moving all around the nail.

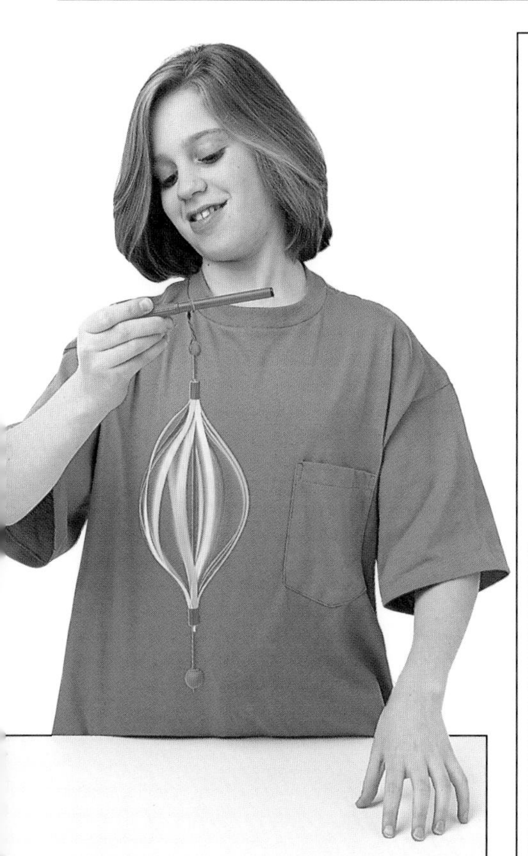

5 Hold the apparatus by the handle, and spin it until the rubber band is completely wound up. Be careful not to bend or crease the strips of paper. Then let it go, and watch "Jupiter's" waistline bulge as the paper "planet" spins around and around.

EXPERIMENT

Jupiter's storms

Jupiter has violent winds and huge storm patterns. The spots we see on the surface are giant eddies lying between streams of gas moving at different speeds. They can alter in size and shape, and sometimes they spin around the planet faster than the planet is actually rotating. Some even alter in color. The Great Red Spot, the largest storm on the planet, was brick-red when it was discovered in 1664, but it is now much paler. Scientists have not yet been able to find the cause for these storms. You can get an idea of what they look like by making "storms" in milk.

You Will Need
- *yellow food coloring*
- *whole milk*
- *red food coloring*
- *clear-glass bowl*
- *dishwashing liquid*

1 Pour about a cup of milk into a bowl. Carefully add one drop each of the red and yellow food colorings. Gently spin the bowl around (not too much) to get the milk moving.

2 Now place one drop of the dishwashing liquid on top of each drop of food coloring. Continue to spin the bowl gently. Then step back and watch the "storm" brewing.

Saturn

SATURN IS FAMOUS for its beautiful rings, which were observed in 1610 by the astronomer Galileo Galilei—though he was confused about what he saw and thought they might be moons. The rings reach so far into space that they would stretch almost from the Earth to the Moon, but they average only a bit more than half a mile (1 km) thick. Saturn is the second largest planet. It is made mainly of hydrogen and helium—much like its neighbor Jupiter (p.72). So far, 18 moons have been discovered, and there may be even more. Most are just chunks of ice, but Titan—the largest of Saturn's family—has an atmosphere that is thicker than the Earth's and contains dense orange clouds.

Saturn's rings
In 1980 a Voyager *spacecraft took this close-up photograph of the rings of Saturn. There are literally thousands of separate narrow rings circling the planet, each made up of millions of icy fragments ranging in size from ice cubes to small cars.*

EXPERIMENT

Saturn's changing shape

Saturn is tilted at an angle of 27°, and the view from Earth depends on the perspective from which we see the planet. This changes as Saturn moves around the Sun once every 29½ Earth years (one Saturnian year). Twice each Saturnian year we see the rings fully open, and twice we see them sideways, making the planet appear ringless. This experiment shows you Saturn's changing shape.

YOU WILL NEED
- *large orange* • *foamcore*
- *compass* • *scissors*
- *poster board*
- *colored pencils*
- *plastic putty*
- *wooden skewer*
- *ruler* • *protractor*

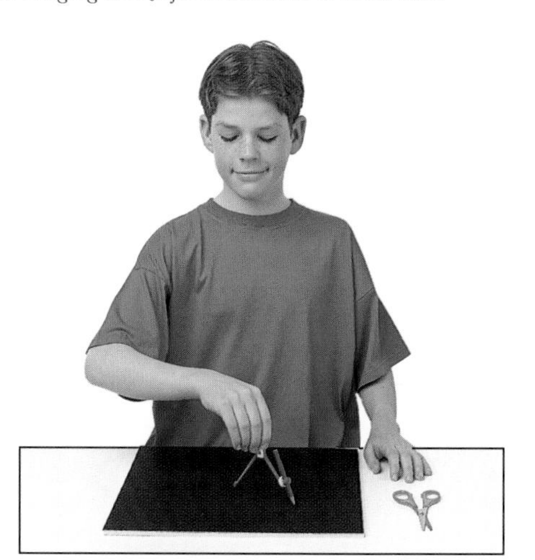

1 USE YOUR COMPASS to draw a 1-ft (300-mm) wide circle on the foamcore, and cut it out. This is the base for this experiment.

2 DRAW A 6-IN (150-mm) circle on poster board. Draw around the orange in the middle of this circle. Cut out this shape. Color it to look like Saturn's rings.

3 PUT SOME PUTTY in the middle of the base. Jab the orange ("Saturn") with the skewer, and put the other end of the skewer into the putty.

4 ADJUST THE SKEWER so that it tilts by 27° to the base, and place the "rings" around "Saturn." Rotate the base, and watch as "Saturn" changes its shape.

EXPERIMENT

Making Saturn's rings and ringlets

Saturn's rings are made of billions of lumps of ice that are arranged into three major rings, with a wide gap (the Cassini division) between the two outer rings. The rings are composed of thousands of tiny rings, nestled one inside the other. These divisions are caused by tiny moons whose gravity diverts the pieces of ice revolving around the planet. You can experiment with making rings below using talcum powder to represent the ice and poster board, a pencil, and a toothpick to represent Saturn's moons.

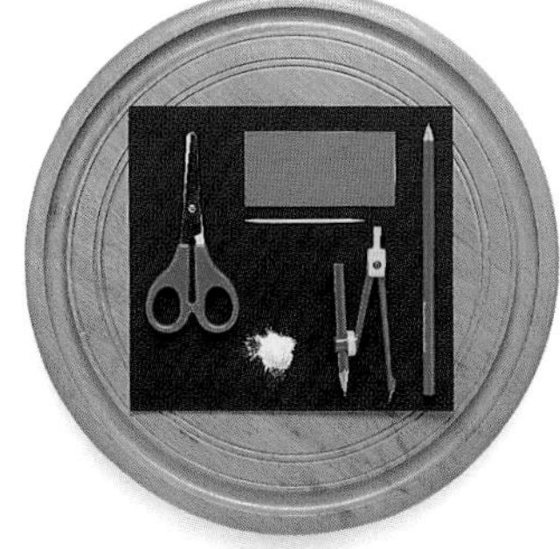

You Will Need

- *rotating cake stand* • *poster board*
- *compass* • *scissors* • *pencil*
- *talcum powder* • *toothpick*

1 Cut enough poster board to cover the cake stand. Evenly scatter talcum powder over this, and sweep in the loose particles at the edges with poster board as the stand rotates. This simulates the effect of the moon Atlas.

2 With the stand rotating, lower a toothpick into the powder at various positions to make tiny "rings." Do the same with a pencil, two-thirds of the way from the center of the stand. This makes the "Cassini division."

Christiaan Huygens

In the 1650's the Dutch astronomer, physicist, and mathematician Christiaan Huygens (1629–95) built a powerful telescope that revealed why Saturn seemed to change shape. This phenomenon had long been a puzzle to astronomers. Huygens reported that the planet "is surrounded by a thin, flat ring, nowhere touching." This was the first discovery of a ring around a planet. Huygens also discovered Saturn's largest moon, Titan, as well as the Orion Nebula. A European space probe to Saturn has been named *Huygens* in honor of the astronomer. Huygens also made significant contributions in other areas of science: he did important research into the nature of light and invented one of the first clocks to have a pendulum—the grandfather clock.

EXPERIMENT

Floating Saturn

Because of its gaseous composition, Saturn is extremely lightweight for its size, and it has the lowest density of all the planets. If there were an ocean large enough to hold it, you would find that Saturn would float. All the other planets would sink because they are denser than water, but Saturn—with a density only 70 percent that of water—is the least substantial planet in the Solar System. Here you can float a mini "Saturn" in water.

You Will Need

- *bowl* • *water* • *marble* • *plastic ball*

1 Fill a bowl halfway with water. Place the plastic ball ("Saturn") in the water.

2 Now place the marble ("Earth") in the water. What happens to "Earth"?

Uranus

URANUS, THE SEVENTH PLANET from the Sun in the Solar System, is extremely cold. It consists mainly of water, with no solid surface and only a few clouds. Unlike any other planet, Uranus is tilted on its side, with its 10 narrow, dark rings and 15 moons orbiting above its equator. Each pole has a continuous period of 42 Earth years in sunshine and an equal period of total darkness. Although it is a giant planet, Uranus lies so far away from the Earth that we can barely see it without a telescope. It was discovered in 1781 by an amateur astronomer, William Herschel, but astronomers knew very little about the planet until the *Voyager 2* space probe swept past it in 1986. *Voyager 2* discovered 10 of Uranus's moons and photographed the 5 largest moons in detail. Its cameras showed that the rings look much brighter when they are backlit by the Sun.

Miranda

Uranus's small moon Miranda is one of the oddest worlds in the Solar System. Its surface is marked with arrow shapes and oval patterns that look like racetracks. Miranda has a towering cliff higher than Mount Everest—a cliff to the same scale on the Earth would reach as high as the orbit of the space shuttle. Some astronomers believe that Miranda was smashed to pieces by a meteorite and then reassembled in space. Another theory is that Miranda contained big chunks of ice that started to melt, producing oval patterns as the water welled upward.

EXPERIMENT

Topsy-turvy planet

Uranus's axis is inclined at an angle of nearly 98° to the vertical, so the planet travels around the Sun on its side. This has an unusual effect on Uranus's seasons as it moves around the Sun on a year lasting 84 Earth years. You can observe this by making a model of Uranus that revolves around the Sun.

YOU WILL NEED

- *foamcore* • *compass* • *scissors*
- *stiff wire* • *protractor* • *orange* • *light bulb*
- *electrical wires and clips*
- *battery* • *screwdriver* • *pencil*

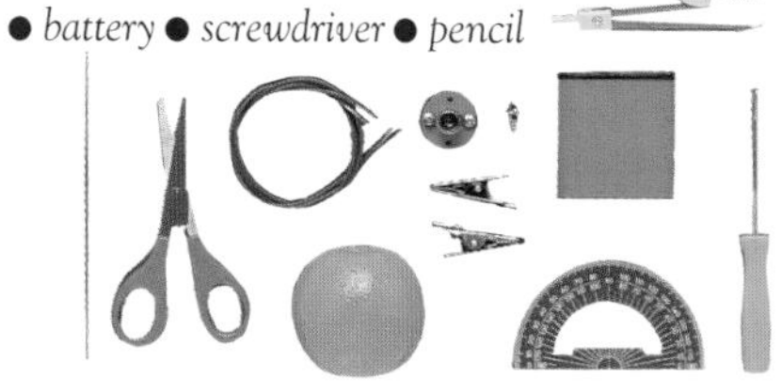

Adult supervision is advised for this experiment.

1 DRAW A LARGE CIRCLE on foamcore, and cut it out. Bend some wire about 8 in (20 cm) long to a 98° angle, and use a protractor to check your accuracy.

2 MAKE A HOLE in the circle's center, attach the bulb, and connect it to the battery. Stick one end of the wire into the orange and the other in the circle's edge.

3 SWITCH ON THE BULB. Rotate the foamcore circle to move "Uranus" on its orbit around the "Sun," while keeping the axis pointing in the same direction.

EXPERIMENT

The discovery of Uranus's rings

The rings of Uranus are so black that you cannot see them from Earth. They were found in 1977 when Uranus passed in front of a distant star. Just before Uranus hid the star and just after it passed the star, the starlight flickered off and on several times. The flickers were caused by the rings. You can observe this effect using your own set of rings.

YOU WILL NEED

- *up to 13 pencils*
- *foamcore*
- *pen*
- *black paint*
- *paintbrush*
- *ruler*
- *flashlight*

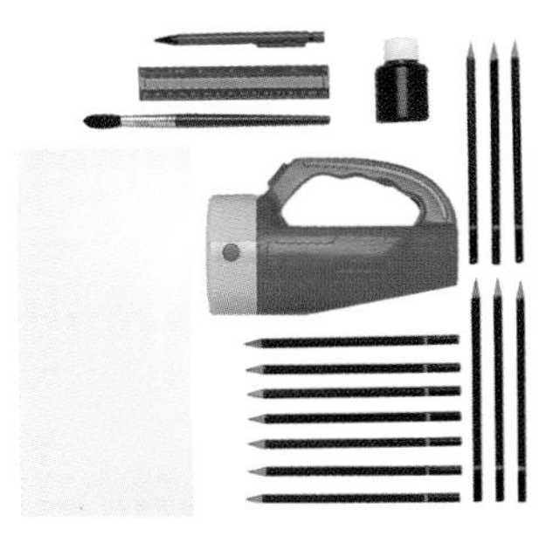

1 PAINT THE foamcore and the pencils with black paint, so that they will not show in the dark.

2 STICK THE PENCILS upright into the foamcore to represent the rings of Uranus. Place a flashlight—the "star"—about 1 yd (1 m) in front of the "rings."

3 DARKEN the room, turn on the flashlight, and move the "rings" from side to side. Observe how they block out the "starlight." When one astronomical object blocks our view of another, it is called an occultation.

■ DISCOVERY ■

William Herschel

William Herschel (1738–1822) was a German professional musician and composer. In 1781, while working in Bath, England, he discovered a greenish blob through his homemade telescope. He thought it was a comet, but it turned out to be a planet, later called Uranus. King George III created a special post for Herschel, who was able to devote the rest of his life to his hobby, astronomy. He went on to identify thousands of double stars and nebulae (shining clouds of gas in space) and worked out the shape made up by the stars of the Milky Way.

EXPERIMENT

Backlit rings

Because the tiny dust particles in Uranus's rings scatter light away from the Sun, instead of reflecting it back to Earth, the rings are most easily seen when they are lit from behind. You can test this using smoke to represent the rings. Ask an adult to light a candle, blow it out, and capture the smoke in a bottle. Then seal the bottle.

YOU WILL NEED

- *clear bottle with cap*
- *matches and candle*
- *flashlight*

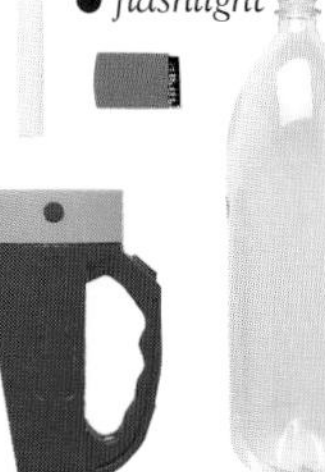

Seeing smoke

Shine a flashlight at the bottle. Can you see smoke particles? Shine the flashlight at the bottle from behind. What happens?

Adult supervision is advised for this experiment.

Neptune

NEPTUNE, discovered by a cunning piece of detective work in 1846, is so far from the Sun that you need a telescope to see its small blue-green disk. Neptune looks like Uranus, and until *Voyager 2* arrived there in 1989 astronomers regarded the two as "twin planets." Neptune, however, is much more exciting than its rather bland neighbor. It has dramatic, changing weather patterns, including a gigantic storm—the Great Dark Spot. Violent winds sweep its sky-blue cloud tops. Its interior is much warmer than that of Uranus, and the planet is circled by four patchy rings. Triton, one of its moons, is the coldest place in the Solar System.

EXPERIMENT

The Scooter and the Great Dark Spot

Pictures of Neptune show a large dark oval (the Great Dark Spot) and a small white spot (the Scooter). As the planet rotates, the Dark Spot comes around every 18 hours, and the speedier Scooter every 16 hours. But the Scooter is actually fixed at the same place on Neptune. The Great Dark Spot is moving, swept along by strong winds that blow in the opposite direction to the planet's rotation, so the Dark Spot lags behind.

YOU WILL NEED

- *revolving cake stand*
- *poster board*
- *modeling clay*
- *tape*
- *compass*
- *scissors*
- *clock with a second hand*

1 PLACE THE CLOCK (representing Neptune) face up on the revolving cake stand. If necessary, you can place a blob of modeling clay underneath the clock to stop it from sliding.

2 CUT a circle of poster board slightly smaller than the reach of the second hand. Tape it over the center of the clock face. The moving tip of the second hand represents the Great Dark Spot.

3 PLACE A BLOB of clay (the "Scooter") on the poster-board circle near its edge. This does not move, while the second hand does—like the winds on Neptune blowing the Great Dark Spot.

4 PUT SOME modeling clay next to the cake stand as a marker. Turn the cake stand counterclockwise at one rotation every 10 seconds. Watch the marker as the "Scooter" and the "Great Dark Spot" pass. Which one comes around soonest?

EXPERIMENT

Pinpointing Neptune

Neptune was originally pinpointed not with a telescope, like its neighbors Uranus and Pluto, but with the aid of mathematics. In the mid-19th century, astronomers became increasingly concerned that Uranus was not circling the Sun in a smooth and regular manner—it seemed that the planet was being pulled out of position by the gravity of a massive world that lay beyond. This world was called Neptune by the astronomers who discovered it. In this experiment you can make your own hidden planet "Neptune" to pull "Uranus" as it moves along its orbit, by using the force of magnetism in place of the force of gravity.

You Will Need

- *1 x 1 ft (30 x 30 cm) poster board*
- *paper fastener* ● *small piece of poster board*
- *paper clip* ● *bar magnet*
- *tape* ● *scissors*

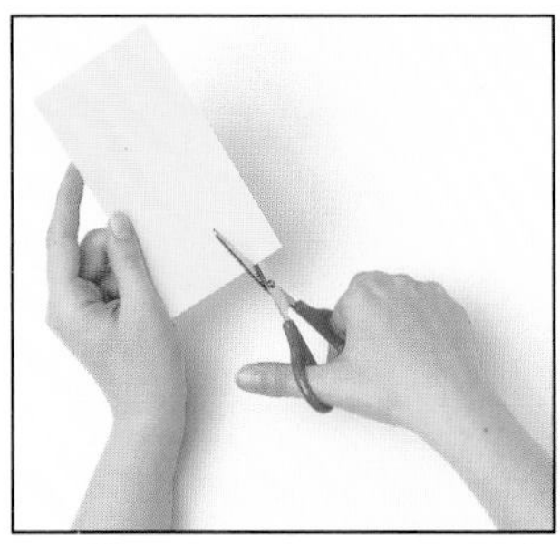

1 Cut the small piece of poster board into a 6 x 1 in (150 x 25 mm) strip. Fold the strip in half, then the halves in half. The bends will allow the board to stretch when you use the magnet.

2 Attach one end of the strip to the center of the large poster board with a paper fastener. The middle fold in the strip should stick up. Make sure that the strip turns easily when you push it.

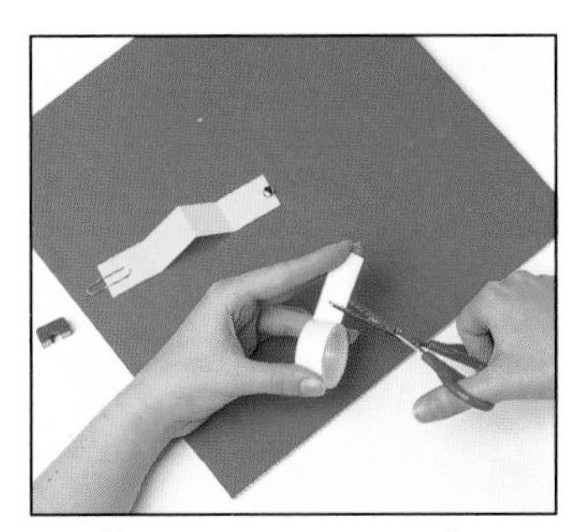

3 Fasten the paper clip ("Uranus") to the free end of the strip, and tape the bar magnet ("Neptune") to the underside of the poster board, near one edge—just outside the "orbit" of the paper clip.

4 Now, turn the strip slowly around the poster board. Can you feel "Uranus" reacting to the pull of the unseen "Neptune" as the "orbit" of the paper clip passes by the hidden magnet?

EXPERIMENT

The condensing atmosphere on Triton

Triton, one of Neptune's moons, is so cold that its "air" (which is composed of nitrogen) freezes on its surface as solid ice without becoming liquid on the way. You can copy this effect using steam.

Adult supervision is advised for this experiment.

You Will Need

- *modeling clay* ● *string* ● *pan*
- *water* ● *wooden skewer*

1 Form a round ball of modeling clay. This represents Triton. Tie a piece of string around it, and tie the free end of string to a skewer. Put this in the freezer for several hours, until the clay is frozen. This mimics the cold temperatures on Triton.

2 Ask an adult to boil some water in a small pan. Then take "Triton" out of the freezer, hold it by the skewer, and put it in the steam for a few seconds. "Triton" is so cold that the water vapor will condense on its surface as a thin layer of ice.

Pluto and Planet X

PLUTO, THE TINIEST planet, was found in 1930 by a young American astronomer, Clyde Tombaugh, who had spent a year searching to see if there were any planets orbiting the Sun beyond Neptune. Earlier calculations seemed to show that both Uranus and Neptune were feeling the gravitational tug of a more distant world. But Pluto, it turned out, was too tiny to pull on these giant planets, and Tombaugh's success was really due to painstaking research. In 1978 a moon, Charon, was discovered to be circling Pluto. Some scientists believe there is another planet, even farther from the Sun than Pluto—Planet X.

Wandering world

Pluto's tilted oval orbit differs from those of all the other planets in the Solar System. It crosses the orbit of Neptune, so Pluto is sometimes closer to the Sun than Neptune is (as it is from 1979 to 1999). There is no danger of collision: the two planets are always apart when their paths cross because Pluto's year is exactly $1^1/_2$ times that of Neptune.

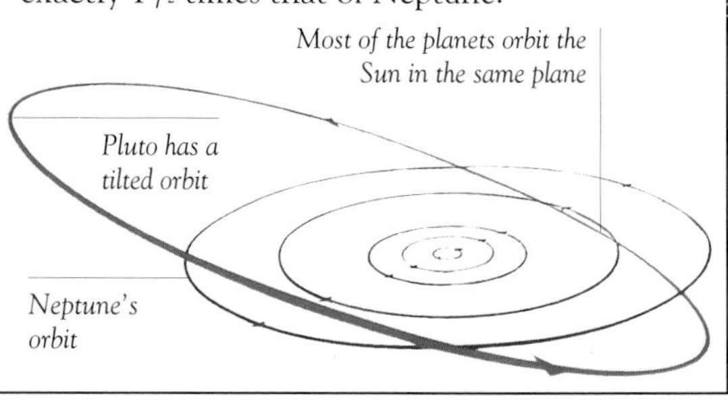

EXPERIMENT

Making a blink comparator

Clyde Tombaugh discovered Pluto using a blink comparator—an instrument that enables you to compare two photographs of the same part of the sky taken at different times. Any small differences between the two photographs can be quickly discovered, even when there are tens of thousands of stars in a photograph. The images are inspected through the same viewing eyepiece and lit one after the other, changing once a second. Any object that is on only one of the photographs seems to "blink" on and off. A planet changes its position between one photograph and the other and looks as if it were jumping backward and forward. Discover your own "planet" by making your own "blink comparator."

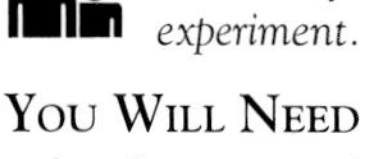

Adult supervision is advised for this experiment.

YOU WILL NEED

- *box (measuring about 8 x 12 in [20 x 30 cm])*
- *2 small sheets of acrylic plastic*
- *tracing paper*
- *pen*
- *thick tape*
- *2 flashlights*
- *ruler*
- *pencil*
- *protractor*
- *glue*
- *scissors*

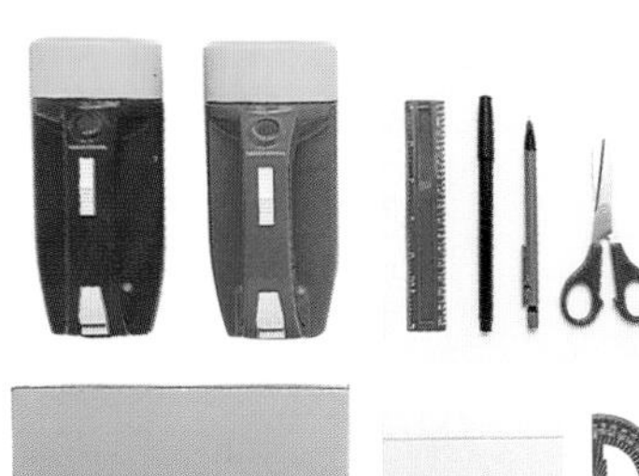

1 CUT TWO SLOTS at 45° to the sides of the box. Cut three holes in the sides of the box. See the diagram, below left, for measurements and positions.

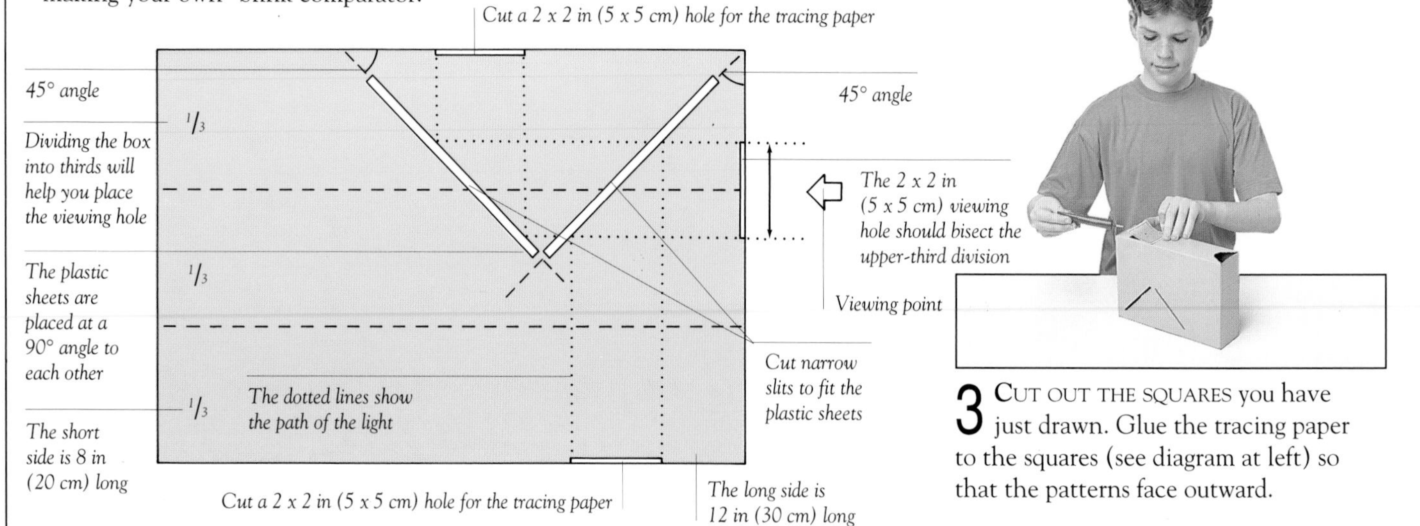

3 CUT OUT THE SQUARES you have just drawn. Glue the tracing paper to the squares (see diagram at left) so that the patterns face outward.

EXPERIMENT

Double planet

A planet and its moon orbit around their barycenter (p.152). Charon, Pluto's moon, is 10 times lighter than Pluto, and the balance point of the two lies outside Pluto. In this experiment you create a similar system.

You Will Need

- *modeling clay*
- *wooden skewer*
- *drinking straw*
- *thread*

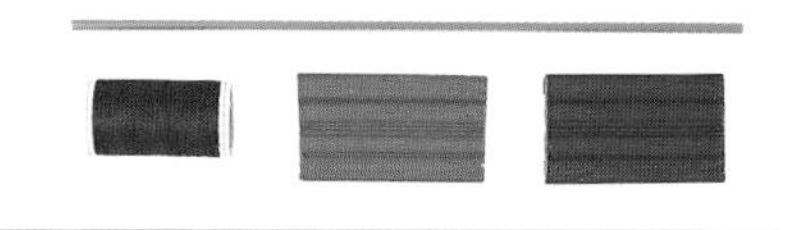

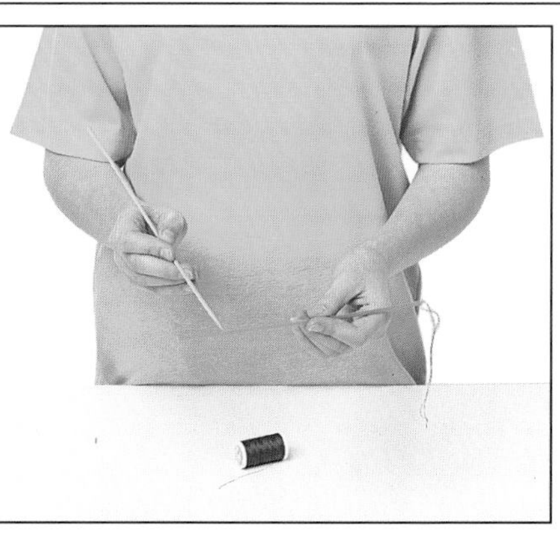

1 Tie some thread to the end of a drinking straw. This is a handle for you to hold. Use the skewer to pierce the other end of the straw, and push the skewer through to its midpoint.

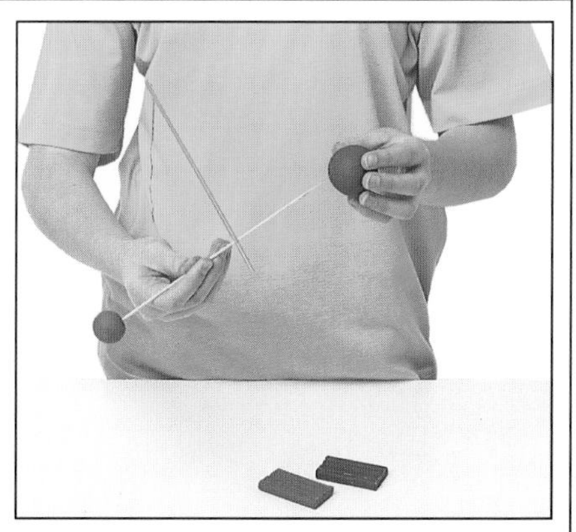

2 Attach two balls of clay—"Charon" 1 in (25 mm) wide and "Pluto" 2 in (50 mm) wide—to the ends of the skewer. Hold the thread, and balance the balls. Push one, so they orbit the barycenter.

Planet X

There may be another planet in our Solar System beyond Pluto. Around 1900 Uranus and Neptune seemed to go "off course," which could have been caused by the gravitational pull of this Planet X. But Uranus and Neptune are now back on track, suggesting that Planet X follows a very long and tilted orbit, which has now taken it away from the other planets. But many astronomers think that the measurements of Uranus and Neptune made around 1900 were inaccurate, and that Planet X does not exist.

The possible orbit of Planet X

The orbit of Pluto

The orbits of the inner planets

The orbit of Uranus

The orbit of Neptune

2 Draw two 2 x 2 in (5 x 5 cm) squares on tracing paper. Add a pattern of stars and a cross to one square. In the other, copy the pattern, but move the cross slightly to one side.

4 Ask an adult to cover the edges of the plastic with thick tape, so that there is no danger of cutting yourself. Insert the plastic into the slits in the box.

5 Shine the flashlights on both sheets of tracing paper. Look through the viewing hole, and make any necessary adjustments to the star patterns so that they are lined up perfectly. With everything in place, look through the viewer again. Ask a friend to turn the flashlights on and off. Can you see any movement?

Adjusting the positions of the plastic sheets very slightly will help you to line up the star patterns

Planetary probes

MUCH OF OUR KNOWLEDGE of the planets has come from robot probes. They have flown past, gone into orbit around, and landed on the planets. A few have been built to withstand the gas plumes and dust in the hearts of comets. While they are all designed to do different tasks, the probes share many common features—for instance, a big dish-shaped antenna that receives instructions from the Earth and sends back data and pictures. Probes can use solar or nuclear power.

EXPERIMENT

Building your own *Galileo* model

Follow these instructions to make a model of the probe NASA sent to Jupiter: *Galileo*. It has detectors, antennas, cameras, a power supply, a propulsion unit, and the *Atmosphere Probe*, which will parachute into Jupiter's atmosphere in 1995.

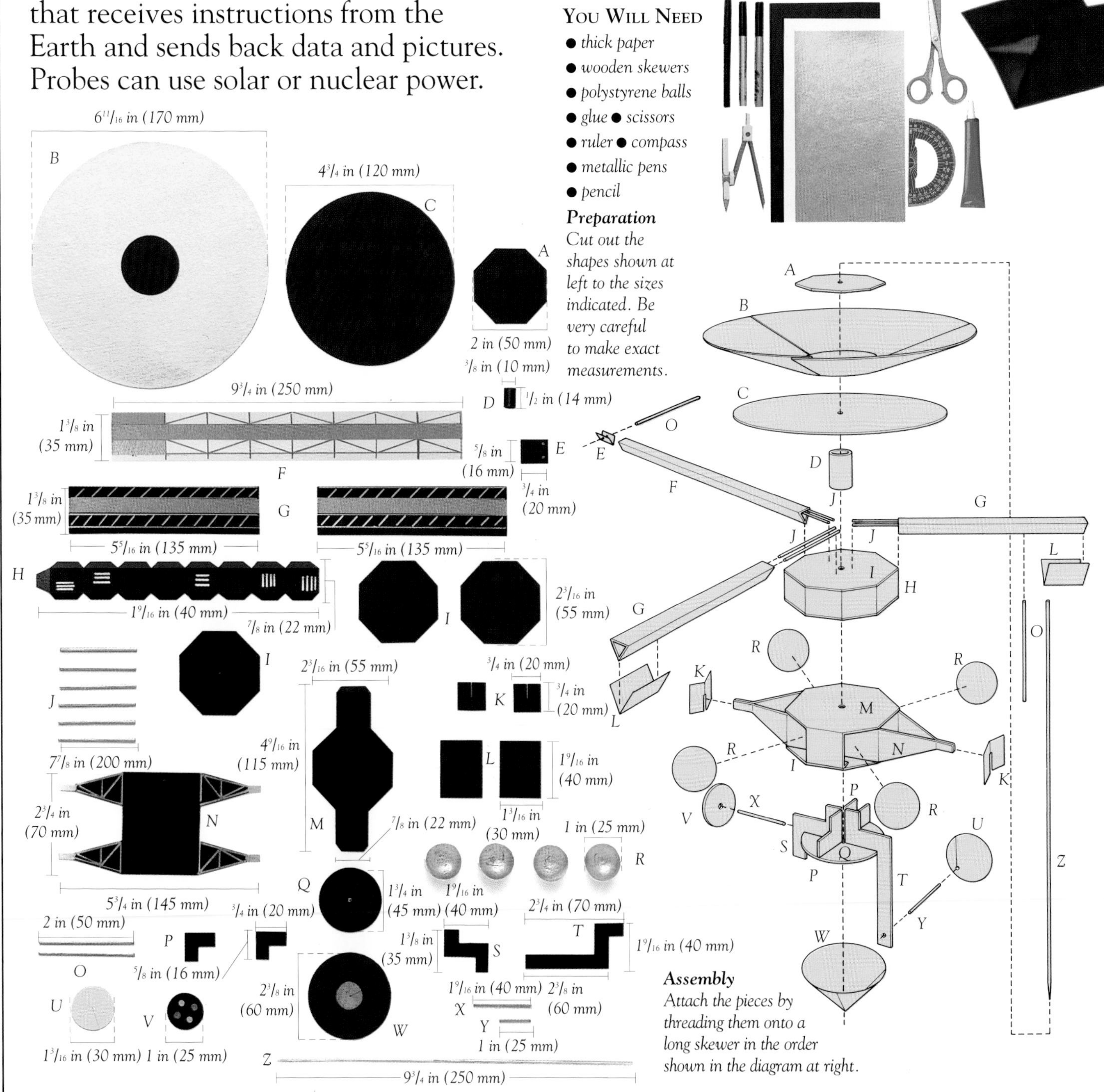

YOU WILL NEED

- *thick paper*
- *wooden skewers*
- *polystyrene balls*
- *glue* ● *scissors*
- *ruler* ● *compass*
- *metallic pens*
- *pencil*

Preparation
Cut out the shapes shown at left to the sizes indicated. Be very careful to make exact measurements.

Assembly
Attach the pieces by threading them onto a long skewer in the order shown in the diagram at right.

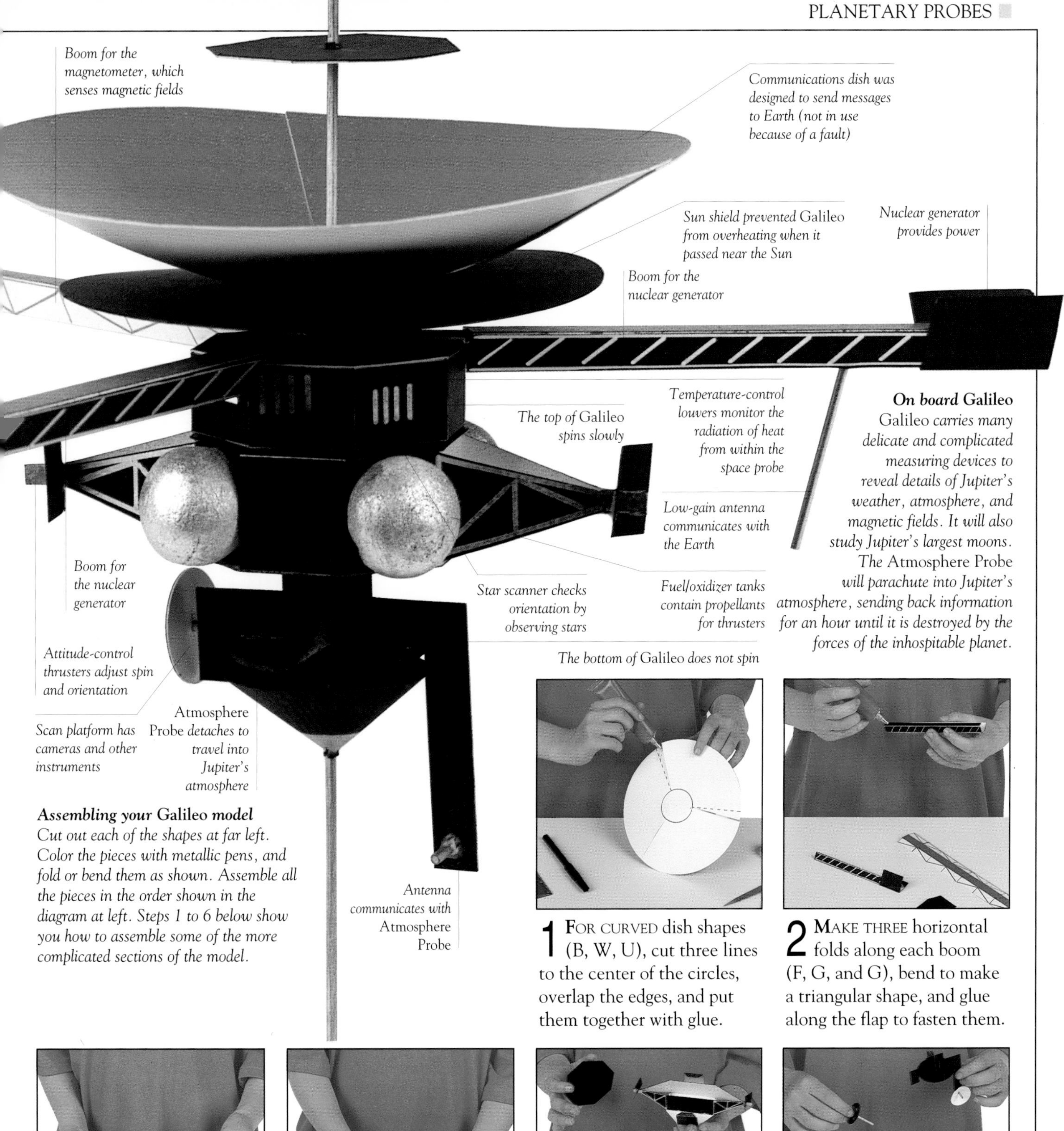

On board Galileo
Galileo carries many delicate and complicated measuring devices to reveal details of Jupiter's weather, atmosphere, and magnetic fields. It will also study Jupiter's largest moons. The Atmosphere Probe will parachute into Jupiter's atmosphere, sending back information for an hour until it is destroyed by the forces of the inhospitable planet.

Assembling your Galileo model
Cut out each of the shapes at far left. Color the pieces with metallic pens, and fold or bend them as shown. Assemble all the pieces in the order shown in the diagram at left. Steps 1 to 6 below show you how to assemble some of the more complicated sections of the model.

1 For curved dish shapes (B, W, U), cut three lines to the center of the circles, overlap the edges, and put them together with glue.

2 Make three horizontal folds along each boom (F, G, and G), bend to make a triangular shape, and glue along the flap to fasten them.

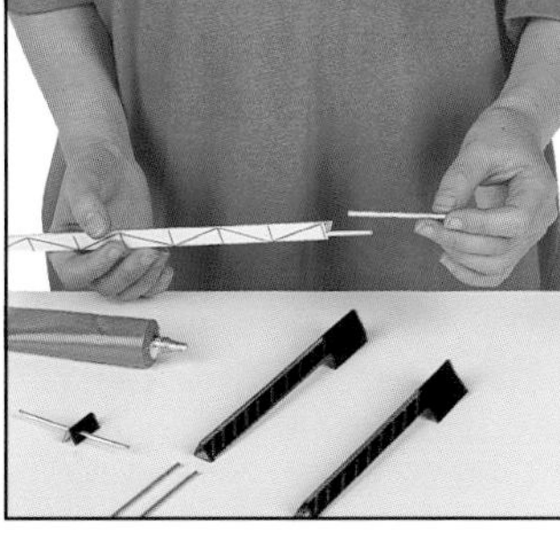

3 Fold two rectangles (L), and glue one to one end of each black boom (G). Glue two skewers (J) to one end of each boom.

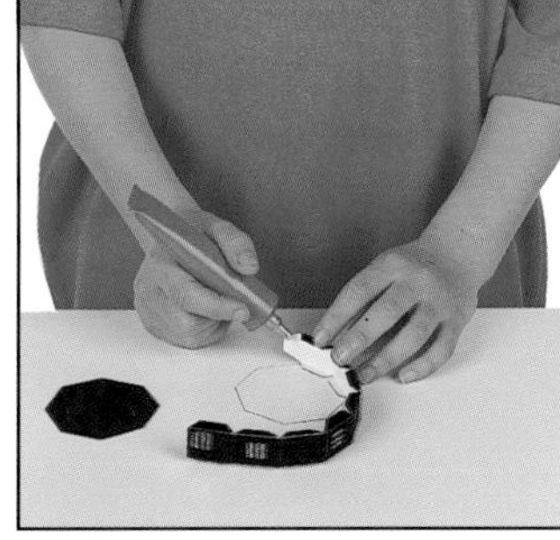

4 Fold the flaps of the body (H), and glue them to the edges of two octagons (I). This will form a box that represents the main body.

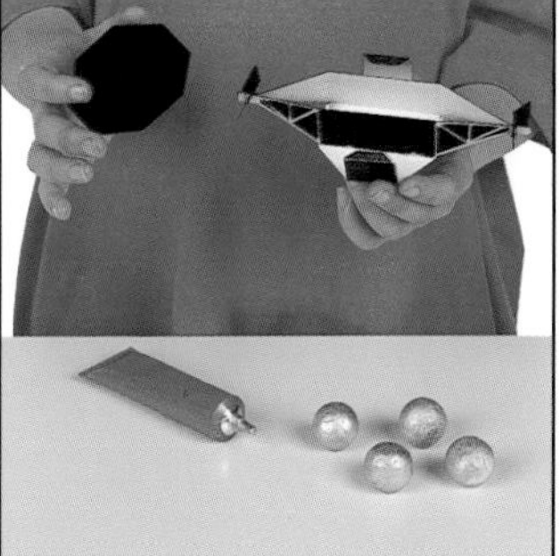

5 Attach the last octagon (I) to the bottom of the "thruster" section (M, N). Glue the four "fuel tanks" (R) to the "thruster" as shown.

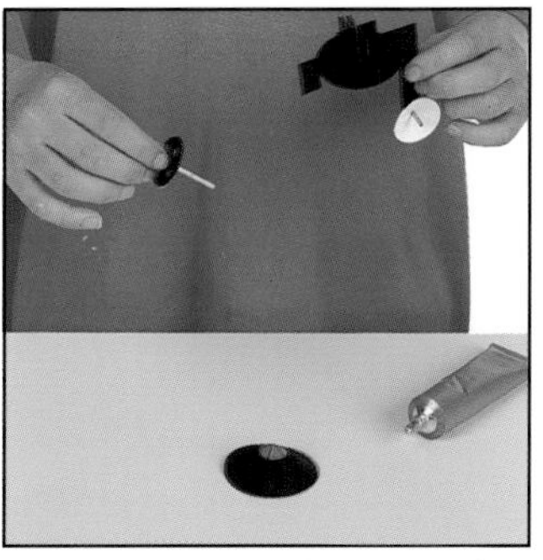

6 Now make the *Atmosphere Probe* model. Attach U and V to small skewers, and slot these into T and S. W slots underneath the "probe."

Comets

COMETS ARE "DIRTY SNOWBALLS" left over from the formation of the Sun and planets some 5 billion years ago. The gravity of a passing star can dislodge comets from their perches in the Oort Cloud (p.154) and Kuiper Belt (p.153). They fall into the Solar System, where they become trapped among the planets. A comet spends most of its life in a frozen state, but as it nears the Sun it rapidly wears out—comets are thought to last only a few hundred-thousand years once they are trapped in the Solar System. Because of the warmth of the Sun, most comets begin to evaporate, developing enormous heads of steam and long tails of glowing gases. Comets move most quickly when they are near the Sun, where its gravity is strongest. In the more remote parts of their orbits, they move more slowly.

Discovering and observing comets

Comets are usually a long way from the Sun when they are discovered, even though they can be very faint. Some astronomers use powerful telescopes to make special searches for comets, which are named after the person who discovers them. Comets are easier to observe as they approach the Sun because they become brighter as they develop heads and, at times, tails. Good times to watch for comets are before sunrise (to the east) and after sunset (to the west).

EXPERIMENT

A comet's orbit

Most comets have orbits that are long and narrow, carrying them close to the Sun and then far away again. At the most distant part of a comet's orbit, it feels the gravity of the Sun only weakly. When the comet is closer to the center of the Solar System, it is affected much more strongly. It must move very quickly to avoid being pulled into the Sun. Below you can see the changing speed of a "comet" as it orbits the "Sun."

YOU WILL NEED

- *foamcore*
- *spoon*
- *paint*
- *paintbrush*
- *marbles*

1 GOUGE A CURVED oval track in the board with a spoon handle—this is the comet's orbit. You can paint the board if you wish.

2 MARK THE position of the Sun just inside the lower part of the track. Tilt the board slightly, with the "Sun" at the bottom. Use a marble to represent the comet. Start it gently along the track from a point at the top opposite the "Sun." Does it travel at the same speed throughout its orbit?

Sun

Comet

■ DISCOVERY ■

Edmond Halley

Edmond Halley was an astronomer and physicist, and the most famous comet of all is named after him. In 1705 he predicted that a comet that had passed the Earth in 1682 would pass again in 1758. He was right, and the comet was named in his honor. The most recent passing of Halley's Comet was 1986. It will pass again in 2062, but this encounter will take place far from the Earth. People will have to wait until 2134 for a good view.

EXPERIMENT

A comet's tail

A comet's tail always points away from the Sun because it is blown back by the force of the solar wind—a stream of high-speed charged particles blowing away from the outer layers of the Sun. Thus, when a comet is traveling away from the Sun, it appears to be going backward. You can make your own "comet" to study this effect.

You Will Need

● *hair dryer* ● *tissue paper* ● *table-tennis ball* ● *tape* ● *scissors* ● *wooden skewer*

1 Cut some tissue paper into about 15 strips measuring ¼ x 8 in (6 x 200 mm). Line them up on the table, and tape them together at one end.

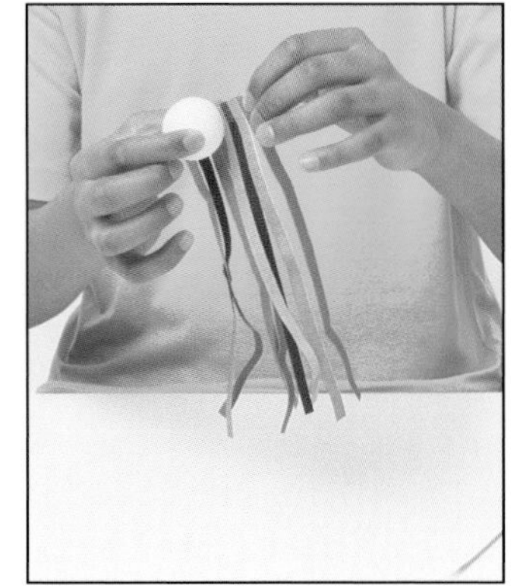

2 Tape the strips (at the taped end) around the middle of a table-tennis ball. Poke the skewer into the ball so that the strips are at right angles to the skewer.

3 Turn on the hair dryer to produce the "solar wind," and point it at the "comet." What happens to the strips? Ask a friend to move the "comet" around you, as you keep the "wind" pointing at it. Do the strips change direction?

EXPERIMENT

Probing a comet

In 1986 when the *Giotto* space probe photographed the nucleus of Halley's Comet under its enormous head of steam, it discovered that the icy core of the comet is coated with some of the darkest material in the Solar System—like a cosmic ice-cream bar. Early in the 21st century, the *Rosetta* space probe will actually land on a comet, drill through the dark surface of the comet, and take a sample for analysis. You can do the same with an ice-cream bar.

You Will Need

● *bowl* ● *warm water* ● *drinking straw* ● *chocolate-coated vanilla ice-cream bar*

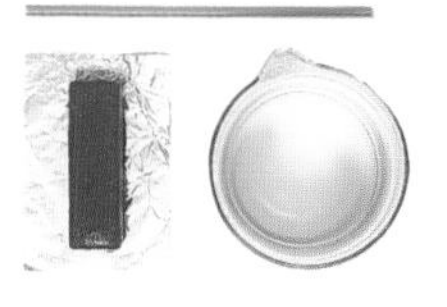

1 Unwrap the ice-cream bar, and set it aside for 15 minutes. Fill a bowl with warm water, and put one end of the straw in the bowl for about 5 minutes.

2 Now, pierce the ice cream with the straw. Dig it in, twist it, and pull the straw out. Blow through the straw to dislodge the "sample" for analysis.

Shooting stars

OCCASIONALLY ON A CLEAR NIGHT you see what looks like a star falling from the sky. A "shooting star," or meteor, actually has nothing to do with the distant stars. It is just a small piece of dust from space that burns up as it falls into the Earth's atmosphere, about 60 miles (100 km) over your head. These dust particles—each about the size of a grain of sand—are shed by old comets. At certain times of the year, the Earth runs into a stream of comet dust, and we see a shower of meteors apparently pouring from the same spot in the sky. These tiny grains of comet dust usually burn up high in our atmosphere. But very rarely a more substantial piece of rock enters the atmosphere and may even survive the fiery descent to land on the Earth as a meteorite.

Catch a falling star
A shooting star, or meteor, here appears to fall near the horizon. The colored lights are the auroras, visible because the photograph was taken near the North Pole. A shooting star lasts only a fraction of a second.

METEOR SHOWERS ***Shower name***	***Date of maximum activity***	***Maximum number of meteors per hour***
Quadrantids	*January 3–4*	50
Lyrids	*April 22*	10
Eta Aquarids	*May 5*	10
Delta Aquarids	*July 31*	25
Perseids	*August 12*	50
Orionids	*October 21*	20
Taurids	*November 8*	10
Leonids	*November 17*	10
Geminids	*December 14*	50
Ursids	*December 22*	15

Where do meteorites come from?
Chunks of rock or metal that fall to Earth are from the asteroid belt, a band of millions of fragments of debris between Mars and Jupiter. This 60-ton chunk of cosmic iron is the largest meteorite known. It fell in prehistoric times near Hoba, Namibia, in Africa.

Meteorite impact

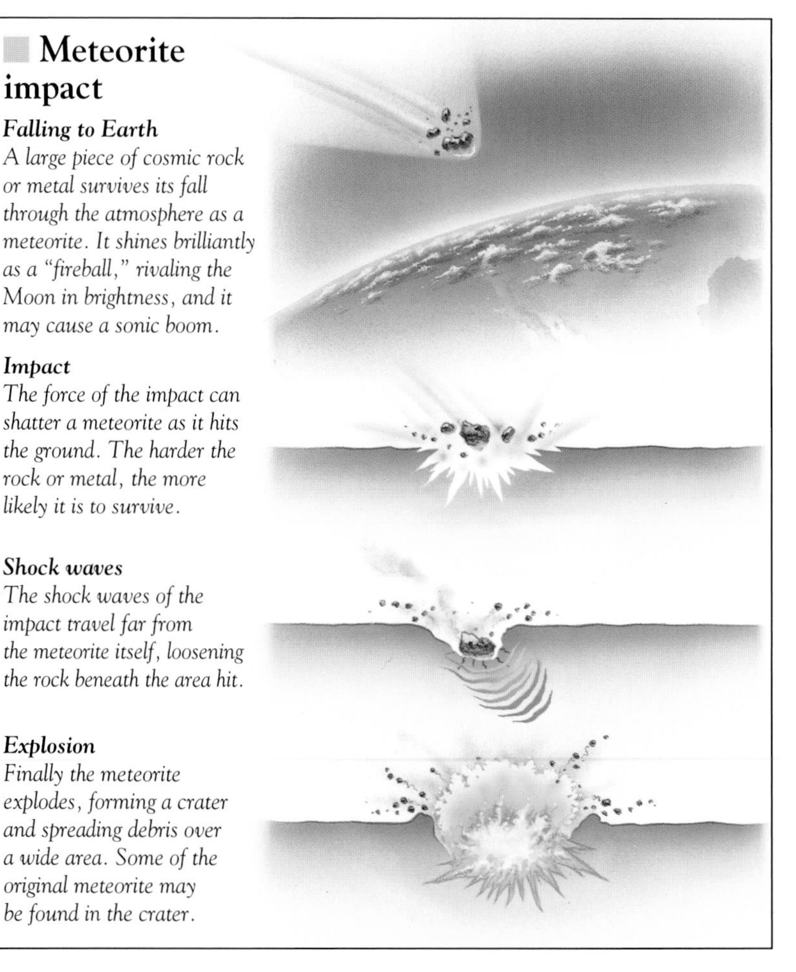

Falling to Earth
A large piece of cosmic rock or metal survives its fall through the atmosphere as a meteorite. It shines brilliantly as a "fireball," rivaling the Moon in brightness, and it may cause a sonic boom.

Impact
The force of the impact can shatter a meteorite as it hits the ground. The harder the rock or metal, the more likely it is to survive.

Shock waves
The shock waves of the impact travel far from the meteorite itself, loosening the rock beneath the area hit.

Explosion
Finally the meteorite explodes, forming a crater and spreading debris over a wide area. Some of the original meteorite may be found in the crater.

Why meteors are hot

When a speck of dust rushes through the Earth's atmosphere, its molecules rub against the gas molecules. This frictional force turns its energy of motion into so much heat that the dust particle burns up and the air glows white-hot. This produces the meteor's shining tail.

Demonstrating friction
You can demonstrate the force of friction by quickly rubbing the palms of your hands together. You will be able to feel the force of friction as it tries to stop your hands from moving. Both hands will become hot because the friction between them is turning the energy of movement into heat energy.

Zodiacal light

The dust shed by comets, which we see as shooting stars when particles streak into our atmosphere, is spread throughout our Solar System. You can see this immense dust cloud at certain times of year when the conditions are right—but you must be in a very dark place with no light pollution (pp.30–31). In spring look to the west after sunset for an eerie, cone-shaped glow with its point high in the sky. This is called the zodiacal light, consisting of countless tiny dust particles lit by the Sun.

EXPERIMENT

Why meteors spread out

Meteors in a shower travel in parallel paths through space, but when they burn up in the atmosphere they seem to spread out from a point in the sky called the radiant. The shower takes its name from the constellation in which its radiant lies. In this experiment, sticks of spaghetti standing parallel to each other suggest the meteor trails.

You Will Need
- *sticks of spaghetti*
- *plastic putty*
- *plate*

1 Evenly spread the plastic putty in a sheet all over the plate. Push a large handful of uncooked, dried spaghetti sticks into the putty, making sure that they do not break.

2 Check that each spaghetti stick is standing up vertically in the putty. The spaghetti sticks represent the particles of dust in a meteor shower entering the Earth's atmosphere.

3 Now look down on the plate. From your viewpoint it will seem as though all the spaghetti sticks were spreading out from the center of the plate. This is what a meteor shower looks like.

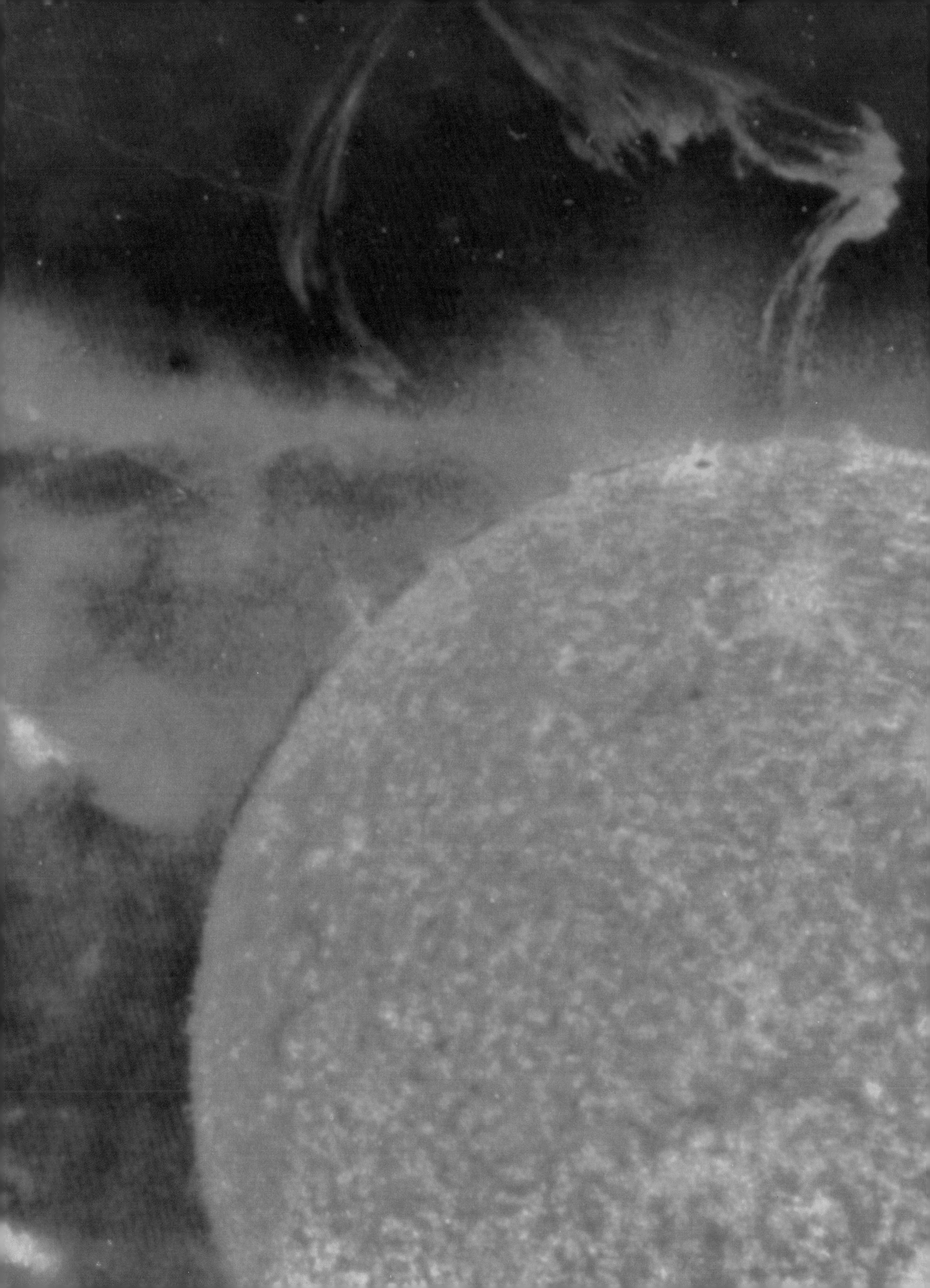

The SUN

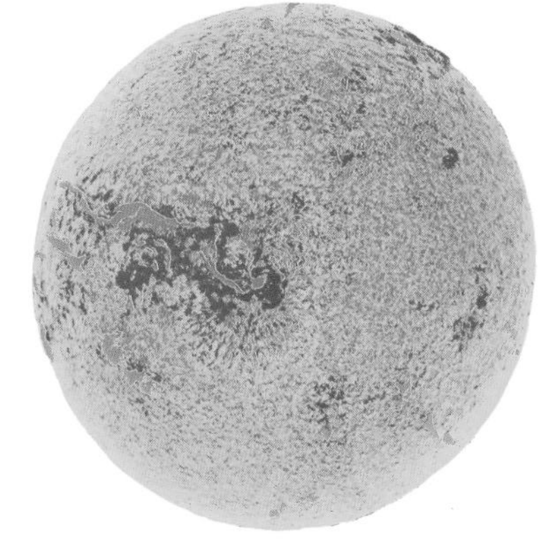

Too hot to handle
Our local star, the Sun, is a seething mass of hot gas. Powerful magnetic fields cause sunspots on the Sun's surface (above) and gigantic glowing loops, called prominences, in the atmosphere (left).

SO VAST THAT A MILLION Earths would fit inside it, the Sun works like a huge slow-motion hydrogen bomb, pouring out energy to its family of planets. The Sun has supplied the Earth with heat and light for the past 5 billion years, and it will continue to provide our energy for another 5 billion years in the future.

OUR LOCAL STAR

THE SUN IS A HUGE BALL OF BLAZING GAS, big enough to contain our planet Earth a million times over. Its brilliantly shining surface, several times hotter than a blast furnace, provides us with both daylight and warmth. Without the Sun's heat, the Earth would freeze to the temperature of outer space, –450° F (–270° C).

The ancient sky
The stars above have not changed for generations. At Stonehenge, in the United Kingdom, an ancient ring of stones has been aligned with the Sun's solstice (p.155)—showing the importance that these events have always held for humans. Ever today, people flock to see the midsummer Sun rise over a special marker in this great stone observatory.

Impressive as it seems to us, the Sun is awesome only because it is very close on the cosmic scale. The other stars in the night sky are all suns too, but they lie millions of times farther away. Deneb, in the constellation Cygnus (the Swan), shines thousands of times brighter than the Sun. Our nearest neighboring star, faint Proxima Centauri, is thousands of times dimmer. If the Earth were orbiting Deneb we would be burned to a crisp; if Proxima Centauri were our Sun we would all be frozen solid.

Studying the Sun

By studying the Sun, astronomers hope to learn how all the stars shine. The Sun, after all, is the only star we can study close up. Almost all the other stars appear only as points of light even through powerful telescopes, though astronomers have been able to make out the rounded shape of some of the biggest and nearest stars. But the very closeness of the Sun makes it dangerous. So much of the Sun's light and heat falls on the Earth that sensitive equipment can be damaged—and that includes human beings, especially our delicate eyes and skin. Never look directly at the Sun, because the heat that will be focused on the back of your eyeball can damage your retina.

In some professional solar observatories, a lens or mirror focuses a magnified image of the Sun on a white screen. In others, the telescope has special mirrors and filters that dim the Sun's light by over a million times, so it is safe to photograph. The simplest way to study the Sun at home is to project its image onto a screen: you can look at the image on the screen quite safely and see details like sunspots—small dark blemishes on the Sun's glowing surface.

The Sun is spotty because it has a patchy magnetic field. Where the magnetic field comes through the Sun's hot surface, it prevents heat from rising. The result is a cooler spot, which looks darker than the rest of the surface. Although astronomers cannot see spots on any other stars, they have good reason to believe that many of the small and cool stars also have them. The light from these stars changes regularly as the stars rotate, and the most likely explanation is that the stars have spots much larger than sunspots. The star's light becomes dimmer whenever its rotation brings a huge dark spot into view.

Seeing our star
This strange structure is a telescope—the McMath Solar Telescope in Arizona. A mirror at the top of the tower reflects light down the diagonal optical tunnel 500 ft (152 m) in length. Deep underground, astronomers study a large focused image of the Sun's disk.

Above the sunspots, loops of the magnetic field protrude into the Sun's lower atmosphere, the chromosphere. Here, magnetism traps gas and controls the Sun's "weather." It also supports huge loops of glowing gas, called prominences. The biggest prominences are 100 times bigger than the Earth. When two magnetic loops touch, they can short-circuit one another. The resulting explosion—a solar flare—can be as powerful as millions of nuclear weapons.

The closest star
Huge pink prominences ring the orange face of the Sun. The furious heat and light of the Sun fuel all life on Earth.

Solar eclipses

Occasionally, the Moon moves in front of the Sun and blocks out the Sun's surface in an eclipse. This reveals the Sun's glowing atmosphere—the corona—extending far into space. In the last century, astronomers who were interested in the Sun's atmosphere had the arduous task of getting to remote parts of the world in order to be in the right place when an eclipse occurred. Occasionally, they managed to

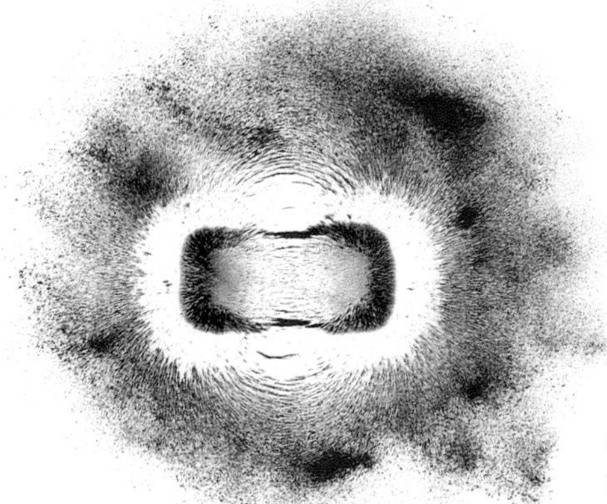

The Sun's magnetic field
The Sun is not only a source of heat and light—it also has powerful magnetic fields, like the magnetism affecting these iron filings. Changes in these fields produce enormous explosions that are called solar flares.

take photographs of the spectrum of the faint corona, which showed unusual bright emission lines (p.153). Some of these came from a new element, helium. To begin with, astronomers thought some other lines came from another new element, which they called "coronium." But they found that these lines actually come from ordinary iron atoms when they are heated to over a million degrees. This showed that the corona is much hotter than the Sun's surface, which is about 9,900° F (5,500° C).

Today, special telescopes (coronagraphs) can show the faint atmosphere at any time. Astronomers can also obtain pictures of the corona from Earth-orbiting satellites that pick up X-rays coming from the corona's superhot gases. The top of the corona is continuously boiling off into space, in a raging gale of hot gases that sweeps out through the Solar System. As this solar wind passes the Earth, some of its particles descend into our atmosphere and light up the air molecules in the colorful display of the auroras: the Northern and the Southern Lights. Over a period of 11 years, the number of sunspots and the strength of the solar wind change. When the magnetism is most powerful and the surface is spottiest, the corona is at its brightest and the solar wind blows most strongly.

By studying other stars very carefully, with special telescopes on the ground and satellites orbiting the Earth, astronomers have picked out radiation very similar to that coming from the Sun's corona. It must be coming from the hot atmospheres of these distant stars. Over a period of a few years, this radiation comes and goes—showing that these stars have a cycle of magnetic activity that is similar to the Sun's cycle.

Inside the Sun

Astronomers have delved more deeply into the Sun's mysteries by splitting up its light into a spectrum of colors, like a rainbow. Each of the colors corresponds to a different wavelength of light. A look at the spectrum of sunlight shows dark lines crossing the colored band. These are produced by different types of atom, each absorbing light at a few particular wavelengths. As a result, without going near the Sun, scientists can work out what it is made of. The Sun is made almost entirely of the two lightest elements, hydrogen and helium. These are in fact the most common substances in the Universe, but the Earth has lost most of its hydrogen and helium because these light gases have largely escaped from the Earth's rather weak gravity. The Sun's central temperature is around 25 million° F (14 million° C). This is so hot that the hydrogen atoms react together, like the ingredients of a hydrogen bomb. This cosmic bomb does not go off all at once, but it undergoes a gradual reaction that produces a continuous supply of energy.

Only in one way does the Sun differ from other stars. We know that it has a system of planets. In fact, many of the other stars probably have planets too, but present-day telescopes are not powerful enough to show them. If we lived on a chilly planet circling the nearby star Proxima Centauri, even the largest planet in our Solar System—Jupiter—would be lost in the Sun's glare. Telescopes on Proxima Centauri's planet would have no hope of picking out the tiny Earth.

Solar discoverer
Sir Norman Lockyer (1836–1920) discovered the element helium in 1868—not on the Earth but in the Sun. He found unknown lines in the Sun's spectrum, which he attributed to a new element. He named it "helium" after the Greek word for the Sun, helios. *Lockyer also studied sunspots and prominences, surveyed Stonehenge, and founded a journal of science.*

Solar eclipse
This photographic sequence shows a total solar eclipse. For a few minutes during the middle of an eclipse the Moon completely covers the yellow Sun, making the solar corona visible—as in the central photograph. This part of the Sun is 1.8 million° F (1 million° C), yet it is not normally bright enough to be seen.

The Sun's energy

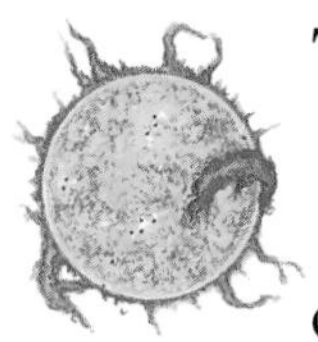

THE ONLY MAJOR SOURCE of energy on Earth that does not derive from the Sun is nuclear energy. All the rest of the energy we use comes from the Sun. We burn "fossil fuels" like coal, oil, and gas to release the energy from sunlight that has been stored in prehistoric plants for millions of years. The electricity we use mostly comes from engines that get their power from fossil fuels. Even wind and wave power depend on atmospheric motions created by the Sun's energy falling on the Earth. Although we are aware that we depend on the Sun for our energy needs, and we know that the Sun's "vital statistics" are huge in every way, it is hard to grasp what its output is in everyday terms. On these two pages, you get an idea of how bright and hot the Sun really is.

Fossilized sunbeams

Almost all sources of light and heat release the Sun's energy from forms where it has been hidden for up to millions of years by layers of rock laid on top. Coal consists of crushed plants that lived 350 million years ago. These plants grew because they could use energy in sunlight to take carbon dioxide gas from the atmosphere and convert it into living cells. Oil and natural gas are the remains of microscopic plants that used the Sun's energy to grow and of microscopic animals that fed on these plants. These simple cells decayed underground and were trapped in rocks, until people sank wells to bring the oil and gas to the surface.

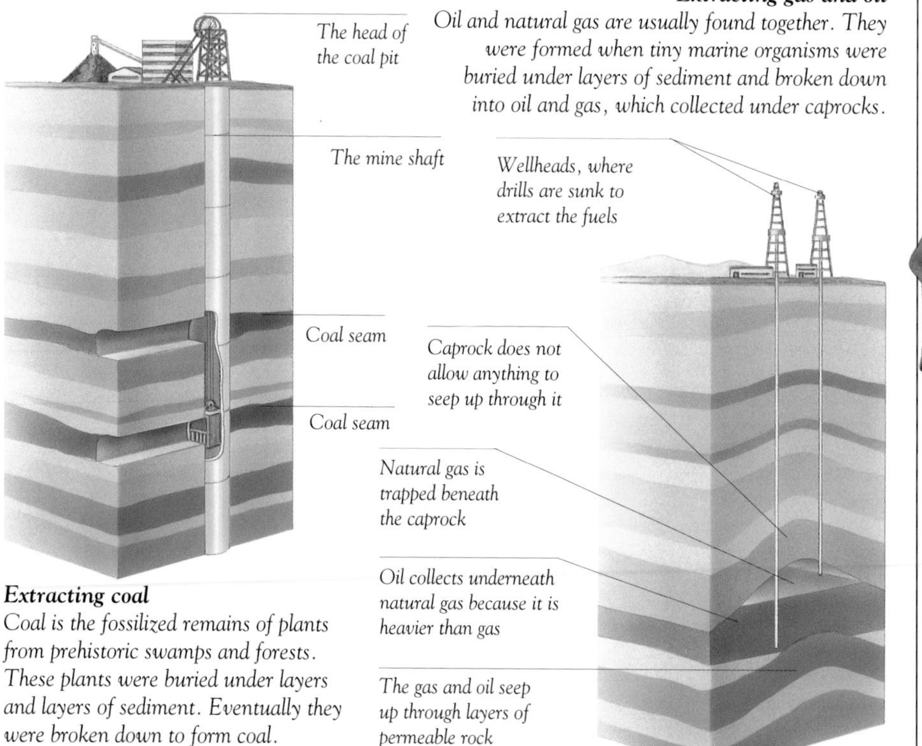

Extracting gas and oil
Oil and natural gas are usually found together. They were formed when tiny marine organisms were buried under layers of sediment and broken down into oil and gas, which collected under caprocks.

Extracting coal
Coal is the fossilized remains of plants from prehistoric swamps and forests. These plants were buried under layers and layers of sediment. Eventually they were broken down to form coal.

EXPERIMENT

The heat of the Sun

Of the Sun's immense output of energy, about half comes out as light and half as heat. We feel the Sun's heat on a warm day. Every square yard of the Earth receives from the Sun as much heat energy as a small electric heater would provide. You can use an ordinary magnifying glass to demonstrate the Sun's heat.

NEVER look at the Sun, because it can damage your eyes.

Adult supervision is advised for this experiment.

YOU WILL NEED
- *magnifying glass*
- *chocolate bar*
- *dish*

Hot chocolate
Take a magnifying glass, dish, and chocolate bar to a sunny spot. Unwrap the chocolate and place it in a dish. Now, use the magnifying glass to focus sunlight on the chocolate. What happens?

EXPERIMENT

How bright is the Sun?

The Sun supplies so much light that it is difficult to imagine just how brilliant it is. In this experiment you can work out how bright the Sun is in terms of something familiar—a domestic light bulb. You need a clear, sunny day and a room that lets the Sun shine in—preferably onto a tabletop where you have a desk lamp.

You Will Need

- *ruler*
- *compass*
- *pencil*
- *notepaper*
- *white poster board*
- *lamp with 60-W bulb*

Electricity from the Sun's heat

These huge mirrors concentrate the Sun's energy at a solar power station in the desert in New Mexico. The concentrated heat is used to boil water, and this drives engines to generate electricity.

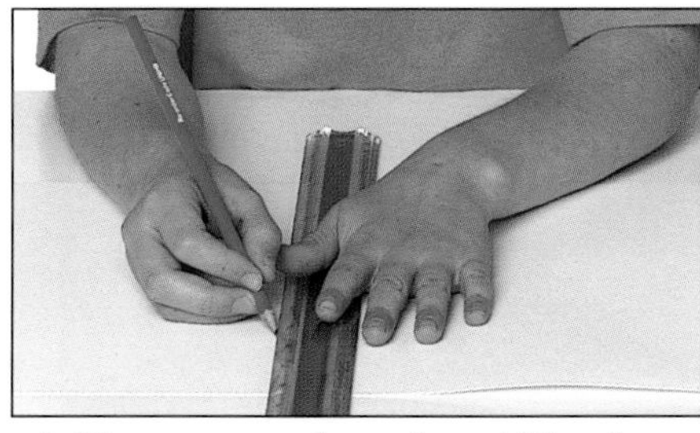

1 Draw a line down the middle of a large sheet of white poster board. Label one half "Sun" and the other "light bulb."

2 Prop up the poster board so that the Sun shines only on the Sun section. Light the light bulb section, and adjust it until the light is as bright as the Sun section. Draw a circle around the lamp light.

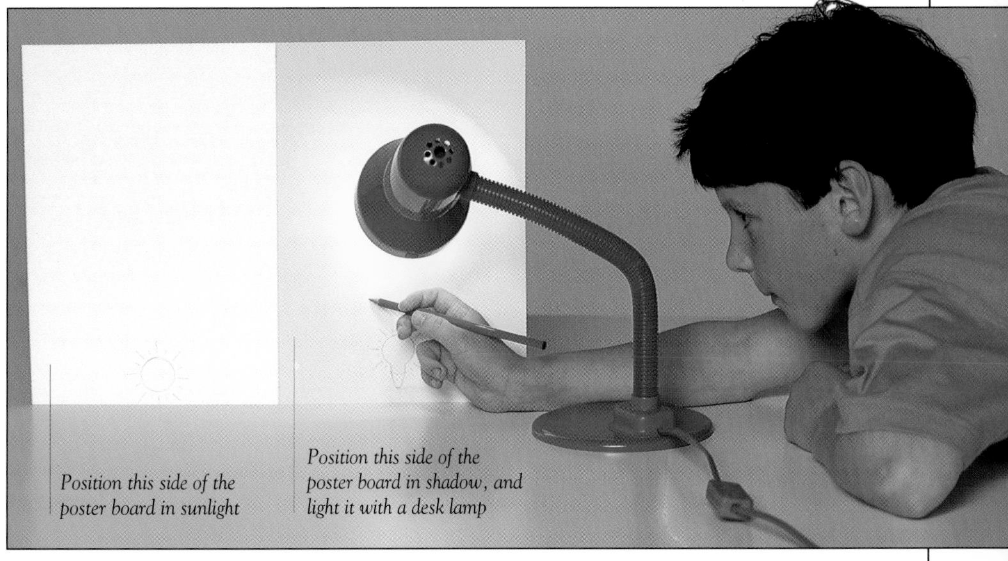

Position this side of the poster board in sunlight

Position this side of the poster board in shadow, and light it with a desk lamp

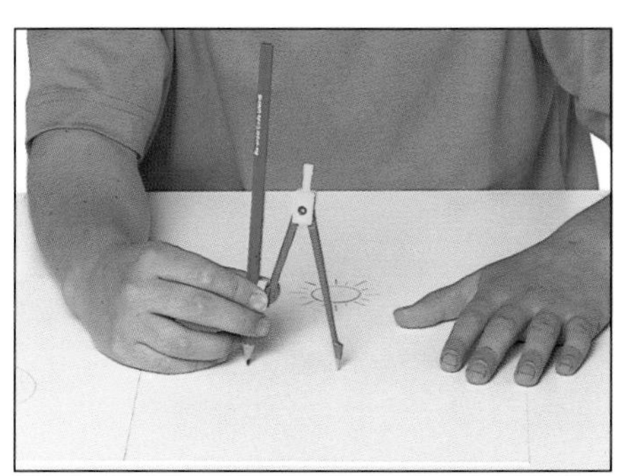

3 Measure the circle, and draw an equal circle in the Sun section. You now know that the Sun is shining on a region the size of your circle with the power of a 60-W light bulb.

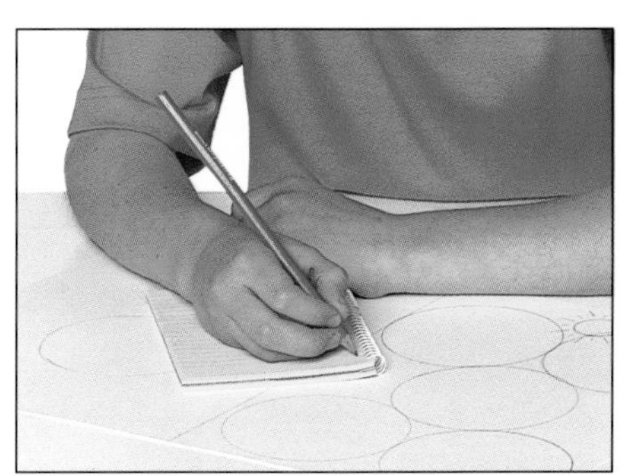

4 Now draw more circles to fill the Sun section completely. Count the number of circles you have drawn. Just on the poster board, the Sun is shining as brightly as this many 60-W light bulbs.

How many 60-W light bulbs?

To calculate the Sun's total output of light:
1. Imagine a sphere around the Sun reaching as far as the Earth. Imagine drawing enough circles to cover it—this is the number of light bulbs equal to the Sun's output.
2. Measure the radius of your circle in inches. Multiply this number by itself (if it is 2, multiply by 2, to get 4). Now, multiply the answer by 3.14 (a number mathematicians call pi). This is the area of your circle in square inches—call this "A."
3. The area of the imaginary sphere is 434 million million million million square inches. Divide this by "A" to find out how brightly the Sun shines, in terms of 60-W light bulbs. The answer: about 50 million million million million light bulbs.

A star close up

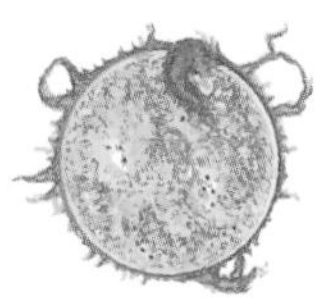

RELATIVELY SPEAKING, the Sun is right on our doorstep. After the Sun, the next nearest star lies more than 270,000 times farther away—which is why almost all stars appear to be just points of light, even in the world's biggest telescopes. But, using safe optical equipment, you can see details on the disk-shaped Sun. What you see is not a solid surface, like that of the Moon, but the topmost layer of a huge mass of gas. The Sun is so hot that nothing on it can exist as solid or liquid —it must be gaseous. You might see dark markings—cool areas on the Sun known as sunspots. Over several days a sunspot will cross the face of the Sun, a result of the Sun's spin.

A simple Sun projector

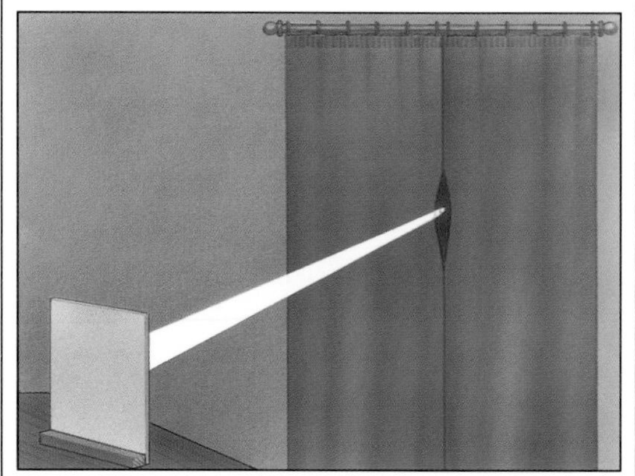

A Sun projector shows the Sun safely. You need a room in which the Sun shines, with heavy curtains. Pierce a 1/16-in (2-mm) hole in poster board. Fix it to the window, near the top, in front of the curtains. Hold a sheet of white poster board in the beam of light to see an image of the Sun on the poster board.

EXPERIMENT

A sunspot projector

The Sun gradually turns around on its axis, carrying the sunspots with it. By watching the sunspots from day to day, you can find out how long it takes the Sun to turn around once. Most sunspots last for many days. Because the Sun does not rotate as a solid body, spots near the equator appear to move around more quickly than spots near the poles. You can make a sunspot projector that will show you many details of our nearest star.

NEVER look at the Sun, because it can damage your eyes.

YOU WILL NEED

- *poster board* • *pen*
- *scissors* • *tape*
- *binoculars*
- *tracing paper*

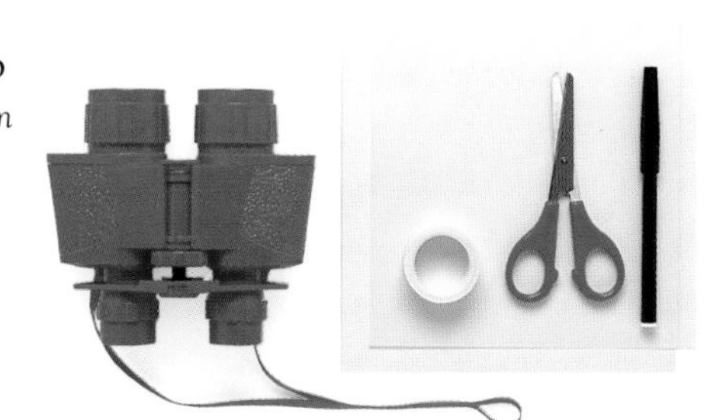

The poster board covers one lens of the binoculars

One lens of the binoculars lets light through, which is then focused on the screen

1 CUT A HOLE just large enough for one of the binoculars' large front lenses, in the center of a sheet of poster board. Firmly tape the binoculars to the poster board, with the hole over one lens. Prop up a second piece of poster board—the screen—facing the Sun. Hold the binoculars about 1 yd (1 m) in front of this screen, so that their shadow falls on it. Slowly tilt and turn the binoculars until a patch of sunlight coming through the binoculars falls on the screen. Focus the binoculars until the image of the Sun is sharp. Prop the binoculars on a chair for extra steadiness.

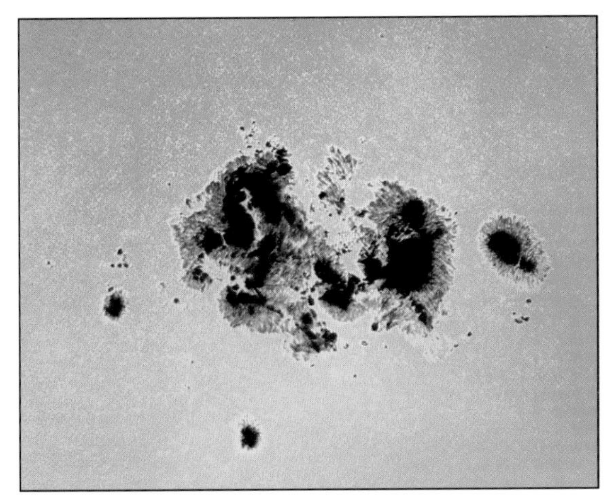

The discovery of sunspots
Sunspots, like the ones shown above, were discovered in the second century A.D. by Chinese astronomers. In A.D. 189 they recorded that "the Sun was orange in color and within it there was a black vapor like a flying magpie."

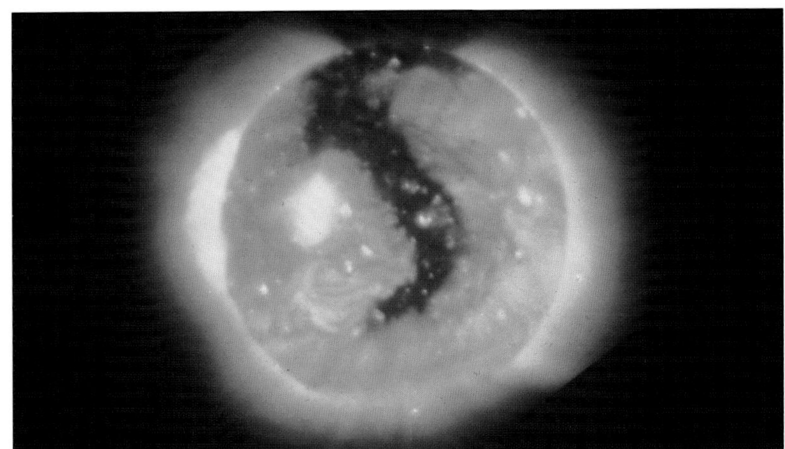

Other ways of looking at the Sun
If your eyes could see X-rays, instead of ordinary light, the Sun would look entirely different. This picture, taken by an X-ray telescope, shows the Sun's atmosphere shining brilliantly, while the surface looks dark. This is because the Sun's surface is much cooler than its atmosphere, which has a temperature of millions of degrees —hot enough to emit X-rays. Glowing patches form above sunspots.

2 TAPE tracing paper to the screen. Draw a circle around the edge of the Sun, and mark any sunspots. Wait a few minutes, and you will see the Sun move sideways as the Earth rotates. Outside the edge of the circle, add an arrow showing the Sun's direction.

3 THE NEXT sunny day, do the same with another sheet of tracing paper. Compare your two drawings, lining them up so that the arrows point the same way. You should be able to identify the same spots, but they will all have moved as the Sun turns.

How big are sunspots?

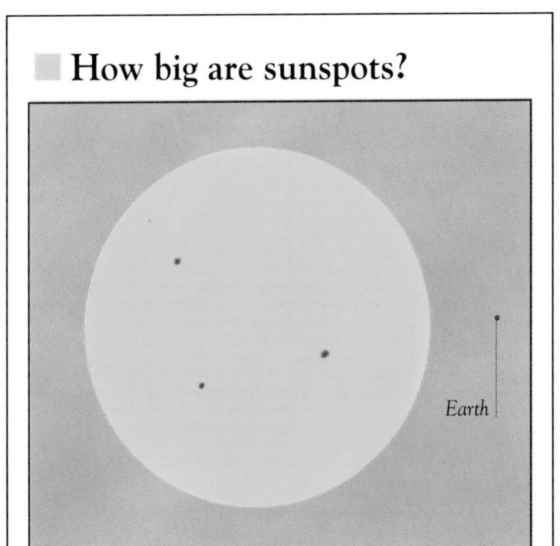

Compared with the Sun, even the biggest sunspots look quite small. The Sun is huge, however, and dwarfs everything else in the Solar System. Once you have recorded some sunspots with your sunspot projector, see how they compare in size with the Earth. Measure the diameter of the Sun as you have drawn it on your screen. Divide the diameter by 100, and draw a blue circle with this new diameter next to the Sun. This is the Earth to the same scale. Compare the size of the biggest sunspot with the size of the Earth. Which is bigger? The largest sunspots ever seen have been 10 times wider than the Earth. Dark spots on other stars—"starspots" — can be so vast that they make the star look dimmer.

The Sun's light

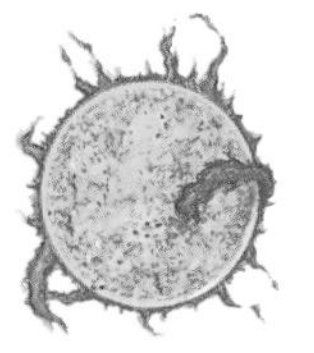

IN 1835 THE FRENCH philosopher Auguste Comte drew up a list of things that would forever remain out of the reach of scientists—including knowing what makes up the stars. Within 30 years, astronomers were routinely studying the composition of the Sun and stars by analyzing their light. Light radiated by the Sun and the stars can be split up with a prism, giving a rainbow of colors—a spectrum—corresponding to light of different energies. This spectrum, which is unique to each star, can reveal its temperature and composition.

Fraunhofer's dark lines

Joseph von Fraunhofer (1787–1826) made the best telescope lenses of his time. One basic problem with lenses is that they do not focus light of different colors equally, which leads to "stray colors" around even the sharpest image. As Fraunhofer was experimenting by passing sunlight through prisms made of his best glass, he was surprised to see that the familiar rainbow was covered with dozens of dark lines. He never discovered why they occurred, but they are named in his memory all the same.

NEVER look at the Sun, because it can damage your eyes.

EXPERIMENT

Making a spectrum

The Sun's white light can be split up into its component colors—its spectrum—in many different ways. Sunlight is split when it travels in and out of a transparent material, such as water or glass, and also when it is reflected off a finely ruled surface. There are many ways to view a spectrum: through a plastic ruler, on a rainbow, in water mist, or on dewdrops. This experiment will show you yet another way.

Rainbow under water
Place a small mirror at an angle in a bowl of water. Use plastic putty to prop up the mirror so that it stays in place. Shine a flashlight down on the mirror, and adjust the flashlight until you see the spectrum on white poster board standing next to the bowl.

YOU WILL NEED
- *flashlight* • *mirror*
- *plastic putty* • *bowl*
- *water* • *poster board*

How a spectrum is produced

White light is bent (refracted) as it passes into a glass prism and bent again as it emerges. The shortest wavelength of light (violet) is bent most, and the longest wavelength (red) least. So the light emerges from the prism fanned out into its constituent colors. White light is produced by hot objects that are solid or very dense, such as the soot particles on a candle flame or the dense gas in the Sun's surface. The rapidly moving atoms in these hot bodies generate light of all wavelengths. But atoms in a thin gas emit or absorb light only at a few very definite wavelengths. Each kind of atom, or element, has a different arrangement of electrons orbiting a small central nucleus. As an electron jumps from one orbit to another, it gains or loses a precise amount of energy. If the electron "steals" this energy from a beam of white light, the result is a dark absorption line at a wavelength corresponding to this energy. When the electron jumps down again, it emits light at this particular wavelength.

White light

Prism

The light is bent when it enters the prism

As the light leaves the prism, it is split into its component colors

Pure white light
Light from a hot solid object consists of all wavelengths. Its spectrum is a continuous "rainbow" containing all colors. The dense gas at the Sun's surface emits light just like a hot solid. If we could observe sunlight at the Sun's surface, it would form a continuous spectrum like this.

Sunlight
The actual spectrum of sunlight is crossed by hundreds of dark Fraunhofer lines—only the most prominent are shown here. These are caused by cooler atoms above the Sun's surface absorbing light at particular wavelengths. Compare these lines with the spectrum to the right. Does the Sun contain sodium and hydrogen?

Sodium light
When white light passes through sodium atoms, they absorb light only in the yellow part of the spectrum. A sensitive spectroscope shows that this dark line is actually double. It is caused by electrons jumping up to two very similar orbits. When sodium atoms are heated, they emit bright yellow light consisting of these two wavelengths.

Hydrogen light
Hydrogen atoms absorb light of several wavelengths. The resulting lines have special names: in the red region is "H-alpha," in the blue-green "H-beta," and in the blue-violet "H-gamma" and "H-delta." Early in this century, the spectrum of hydrogen gave scientists a clue to how electrons orbit in atoms.

EXPERIMENT

Making Fraunhofer lines

An instrument called a spectroscope produces a very detailed spectrum in which you can see all the details—including the bright and dark lines that are the "fingerprints" of elements. Here you can make a simple spectroscope.

Adult supervision is advised for this experiment.

You Will Need

● *magnifying glass* ● *candle* ● *camping stove with gas* ● *teaspoon* ● *table salt* ● *scissors* ● *tape* ● *prism* ● *poster board* ● *paper* ● *foamcore* ● *matches*

1 Cut a 6 x 1 in (150 x 25 mm) section out of foamcore, and cover this with paper to make a slit 6 x 1/16 in (150 x 1 mm). Set up a poster-board screen about 20 in (50 cm) away on which to shine the light.

2 Ask an adult to light a candle and to set it behind the slit. Hold a magnifying glass in front of the slit, and use it to focus the light on the screen. Move the magnifying glass toward the screen until you can see an image of the slit. Adjust the screen to "focus" the spectroscope and make the image sharp.

3 Hold a prism behind the magnifying glass, and turn the prism to make a sharp spectrum on the screen. Now look from the position of the spectrum on the screen into the prism, moving your head from side to side. How many colors can you see?

Creating a new light

To see the spectrum of table salt, use the same setup as above. Ask an adult to replace the candle with a lit camping stove and then to put a pinch of table salt in the flame, making a bright orange-yellow color. Now look into the prism. How many colors do you see? Salt is made of sodium and chloride atoms. Compare your colors with the sodium spectrum above.

Inside the Sun

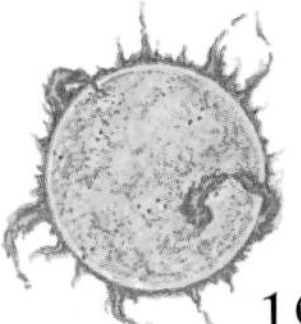

WHAT LIES INSIDE the Sun, and what makes it shine? The answer became clear in the 1930's, when scientists discovered reactions between the nuclei of atoms. The gases in the Sun's core are under pressure from the layers above, and gases under pressure become hot—the core is at 25 million° F (14 million° C). Hydrogen nuclei join together there in a process called nuclear fusion, and this creates energy. The energy starts off as lethal gamma rays and X-rays, but as it spends a million years passing through the overlying gases it becomes safer light and heat. In the upper layers of the Sun, the energy circles in "convection currents"—like swirling eddies of water in a heated pan. On reaching the surface, the energy takes only 8.3 minutes to reach the Earth.

■ A Sun made of coal?

Before astronomers discovered what the Sun was really made of, there were many theories about what made it shine. In the 19th century it was suggested that the Sun might be an enormous lump of burning coal—but even a piece of coal as big as the Sun would burn out relatively quickly. Some people thought the Sun was a world covered with spewing volcanoes, as in the picture below. Astronomers had to wait until the 20th century to discover the truth.

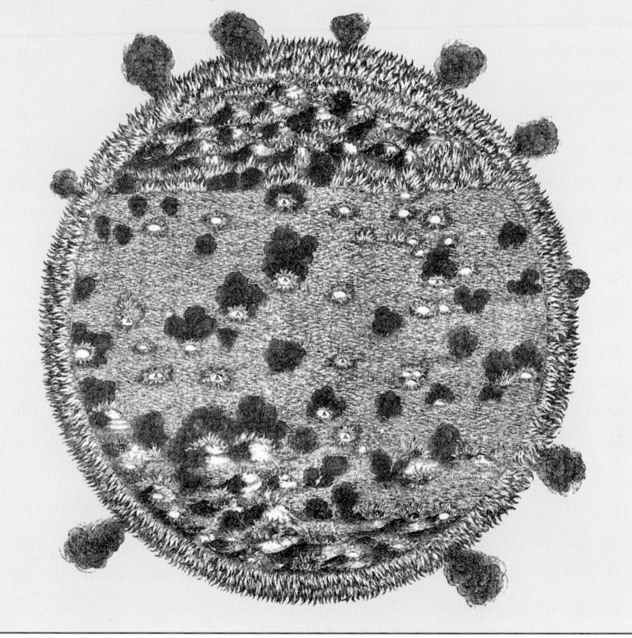

EXPERIMENT

Ringing like a bell

Astronomers cannot see right into the Sun's core, but there are several ways that they can find out what goes on inside. One way is to investigate how the Sun vibrates—like a ringing bell or a glass filled with liquid. While a glass vibrates hundreds of times per second, the Sun is so huge that each vibration takes about 5 minutes. The way it rings gives clues about conditions inside the Sun—such as the density, pressure, and viscosity. Ask a friend to help you demonstrate how different substances in a glass can make different vibrations.

YOU WILL NEED

- *4 identical glasses* ● *spoon*
- *water* ● *cooking oil*

Making music

Pour an equal amount of water into three of the glasses until they are three-quarters full. Pour oil into the fourth glass to the same level. Now, ask a friend to line up the glasses in random order while you turn away. Your friend should tap each glass lightly with a spoon while you listen carefully to the ringing sound. Can you tell which glass contains oil? Why does it sound different?

NEVER look at the Sun, because it can damage your eyes.

The layers of the Sun

The Sun is made up of five main layers. The central region, or core, is where the Sun's energy is produced in the form of X-rays and gamma rays. This radiation passes to the radiative zone, where it loses some of its energy. In the convective zone, the energy swirls back and forth beneath the surface. It then passes through the photosphere and out into space through the Sun's lower atmosphere, the chromosphere.

The photosphere is the Sun's visible surface

In the convective zone, gases circulate in giant eddies

In the radiative zone, radiation from the core loses energy

The core is where nuclear reactions produce all the Sun's energy

Sunspots (p.100) appear on the Sun's surface in cycles

Solar prominences arc from one sunspot to another

The chromosphere is the Sun's lower atmosphere

Convection currents

Heat can travel by radiation, convection, or conduction. Two of these ways occur within the Sun: radiation near the core and convection toward the surface. Convection is the transmission of heat by currents of gas or liquid. It can be demonstrated by adding glitter to hot water.

How the Sun shines

Nuclear fusion works by bonding together the centers, or nuclei, of the Sun's hydrogen atoms. This complex reaction turns four hydrogen nuclei (protons) into one nucleus of helium. A helium nucleus is only 99.3 percent as heavy as four individual hydrogen nuclei, so the extra 0.7 percent is turned into pure energy. Every second, over 4 million tons of the Sun are converted into energy.

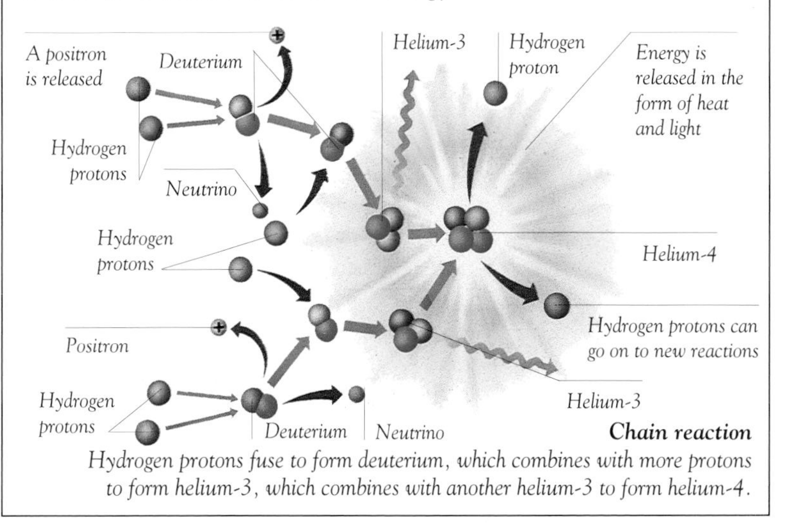

Chain reaction
Hydrogen protons fuse to form deuterium, which combines with more protons to form helium-3, which combines with another helium-3 to form helium-4.

The solar cycle

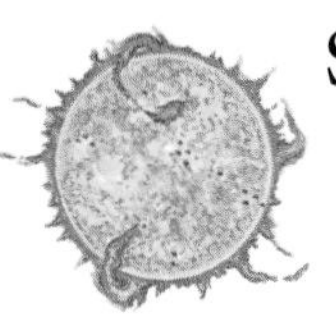

SOMETIMES THE SUN appears to be covered in dark blotches—sunspots. At other times it is almost spot-free. Although sunspots can appear and disappear in weeks, the number of sunspots builds to a maximum roughly every 11 years and then dies away again. This "solar cycle" of sunspots is somehow driven by the magnetic field in the Sun's surface layers. Over the years the field gets "wound up" like a twisted rubber band—possibly because the poles and equator of the Sun spin at different rates. The winding up makes the field stronger, and more sunspots appear. The whole Sun becomes much more active at "sunspot maximum," with effects that can be felt way beyond the Sun—even on Earth. Solar flares are giant explosions near sunspots, and they spew out electrically charged particles that can disrupt TV programs, cause electronically controlled garage doors to open, and even lead to power blackouts.

The mini "ice age"

Some scientists claim that sunspot activity affects the Earth's weather, with cooler weather linked to an absence of spots. The effect is most apparent during the periods when the Sun is free of spots for decades. In 1684 Britain's Astronomer Royal, John Flamsteed, described a particular sunspot as "the only one I have seen in his face since December 1676." Many years later the astronomer E. W. Maunder discovered that the Sun had been virtually spot-free from 1645 to 1715. This "Maunder Minimum" coincided with a bitterly cold period—the mini "ice age"—in northern Europe. In London the River Thames froze, and "frost fairs" were held on its icy surface.

Iron filings react to the force of magnetism

"Arch" of magnetism —a prominence

Pair of magnetic poles—sunspots

Magnetic influences

The magnetic field that causes sunspots can be imagined as two poles of a magnet thrusting up through the Sun's surface. If you held the two ends (the poles) of a horseshoe magnet under this page, then sprinkled iron filings onto the paper, you would see a pattern (left) emerge. The filings cluster around the two magnetic poles, which is why sunspots generally come in pairs. You can see that filings are also attracted to an "arch" of magnetism stretching from one pole to the other. In the Sun's atmosphere, the magnetism can support a glowing arch of hot gas from one spot to another—a prominence. Some prominences hang quietly over the Sun or puff away into space. But if two prominences touch, the result can be a violent explosion. This is known as a solar flare.

EXPERIMENT

The sunspot cycle

The number of spots on the Sun comes and goes over a period of 11 years. The first few spots of a new cycle break out at high latitudes, near the poles of the Sun. Over the next few years, the spots occur closer and closer to the equator and increase in number. After reaching a peak, the number of spots diminishes again, but the spots continue to occur ever closer to the equator. As the final spots of this cycle die out, the first spots of the new cycle begin at high latitudes. By making the model below, you can see several "11-year solar cycles" take place before your eyes in just a few seconds.

You Will Need

- *black and yellow poster board*
- *scissors* • *ruler*
- *compass* • *glue*
- *black marker* • *pencil*

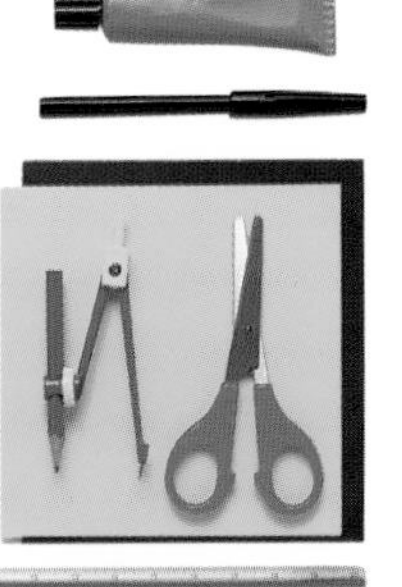

1 Cut out a 12 x 6 in (30 x 15 cm) piece of black poster board. Fold it 1 in (2 cm) from one end, and cut out a circle 5 in (12 cm) in diameter next to the fold. Glue a square of yellow poster board over the hole.

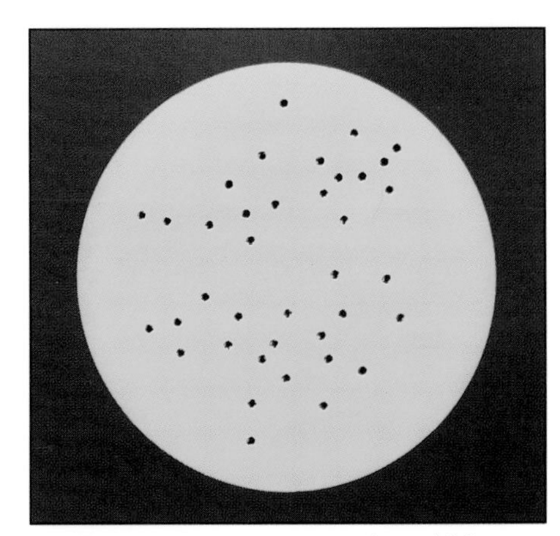

2 Fold the black poster board like a card with the "Sun" on the front. Glue the small fold to the larger one so that the whole is joined up like a sleeve.

3 Turn it over, and with your pencil point, make holes in the yellow "Sun" in the pattern shown above Step 2, with fewest at the "equator" and "poles."

4 Cut a strip of poster board measuring 20 x 6 in (50 x 15 cm), with a tab for pulling. Draw V patterns as shown below, and color the "V's" black.

The "V's" show periods of sunspot activity

Pull to the right

5 Insert the tab end of the strip into the left side of the sleeve (you may have to trim the edges if it sticks), and push it through until the tab is visible on the right. Grasp the tab and pull slowly. Watch how "sunspots" appear near the "poles," then move toward the "equator," reaching a maximum on the way.

Solar eclipse

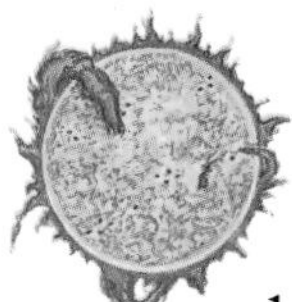

ALTHOUGH THE SUN is 400 times bigger than the Moon, it is also 400 times farther away from the Earth. As a result, the two appear to be the same size in the sky. From our viewpoint on the Earth, the Moon's disk can just overlap the Sun's disk and blot out its light, causing a solar eclipse. An eclipse of the Sun—even a partial one—is rare, and a total eclipse is something you will be lucky to see from your home once in a lifetime. But it gives a stunning view of the Sun's atmosphere. Astronomers travel all over the world to get a good view of a total eclipse. If a solar eclipse is due to be visible in your neighborhood, the newspapers and local radio will let you know.

EXPERIMENT

Observing an eclipse

NEVER look directly at a partial solar eclipse or use binoculars or a telescope. Your local astronomy group may have a public observing session. If you have made a sunspot projector (pp.94–95), you can also project an image of the eclipsed Sun. If not, here is a purpose-built eclipse projector.

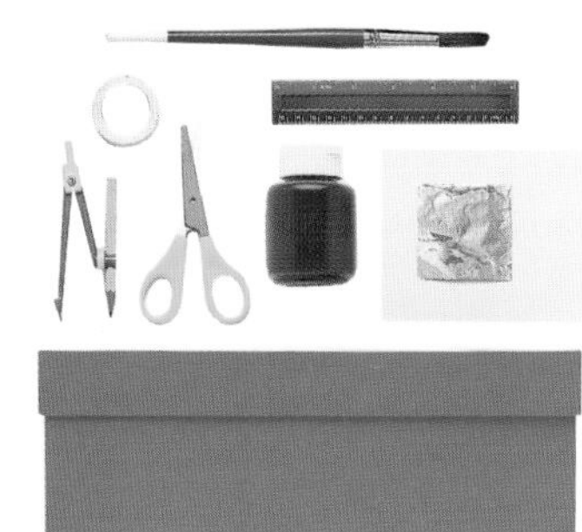

YOU WILL NEED

- *shoe box*
- *paintbrush*
- *ruler* • *tape*
- *tracing paper*
- *scissors*
- *aluminum foil*
- *compass*
- *black paint*

1 CHOOSE A SHOE BOX with a lid that fits well, and paint it black inside. For the best results, the box must not allow any light inside, like a camera.

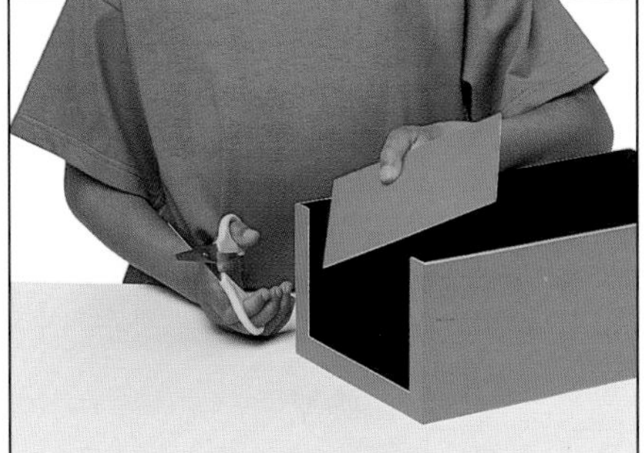

2 CUT OUT ONE END, leaving an edge of about 1 in (25mm) on each side. In the center of the other end, cut a circular hole about 1 in (25 mm) wide.

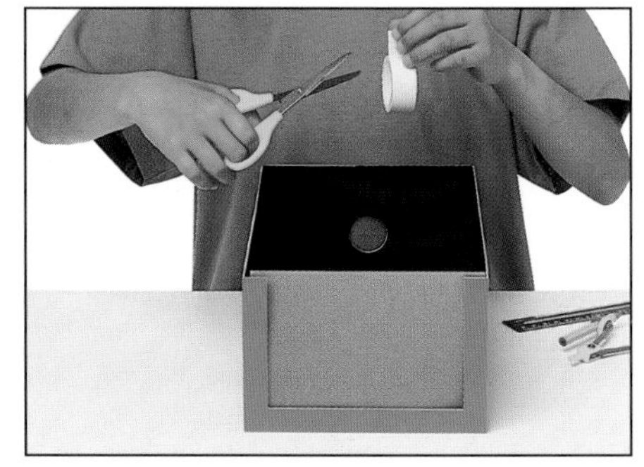

3 COVER THE END of the box you have cut away with a "screen" of tracing paper. Attach the paper with tape all around its edges inside the box.

4 TAPE A SMALL PIECE of aluminum foil across the circular hole at the other end of the box. Make sure the foil is stretched tightly. With the compass point, carefully pierce the foil to make a very small hole (a pinhole) in its center. If you tear the foil even a little, replace it and start again.

NEVER look at the Sun, because it can damage your eyes.

5 WHEN THE ECLIPSE is about to begin, hold the box so that the pinhole end points toward the Sun. A tiny image of the Sun will appear on the paper screen at the other end. Watch the shape of the Sun change very gradually as the Moon covers up part of the Sun's bright disk. Keep a record of the eclipse by sketching it every 10 minutes.

EXPERIMENT

Making a solar eclipse

You can imitate a total eclipse. Look through a hole in the center of the shadow (the umbra) cast by your "Moon," and the "Sun" should be blocked out. By looking through holes at the edges of the shadow (the penumbra), you can see the "Moon" only partly covering the "Sun." This imitates a partial solar eclipse, which is visible from a larger region of Earth than the total eclipse. You can see how an eclipse looks from space.

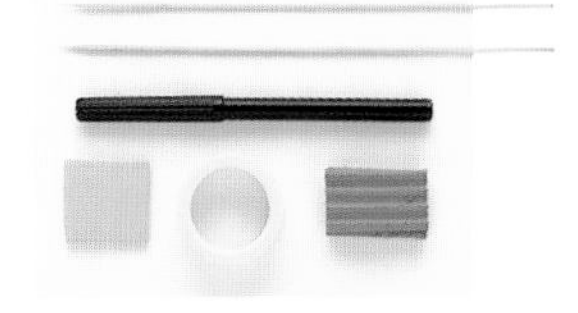

YOU WILL NEED

- *modeling clay*
- *pen*
- *2 wooden skewers*
- *plastic putty*
- *lamp (without shade) with 40-W bulb*
- *thin white poster board*
- *ruler*

1 MAKE A BALL of modeling clay about 1 in (25 mm) wide to represent the Moon. Stick it on top of the skewer, and push the other end of the skewer into a lump of plastic putty to act as a base.

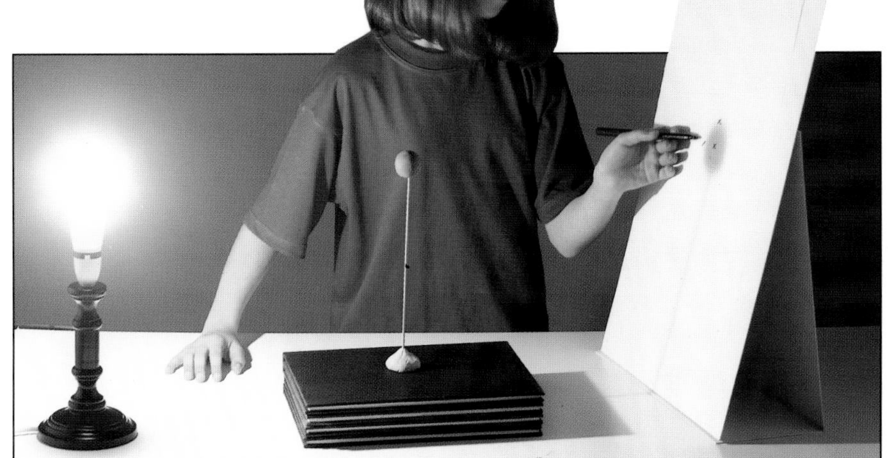

2 PLACE THE LAMP (the "Sun") and the "Moon" so that they are at the same height. With the light off, place a poster-board screen where your eye would need to be to see the "Moon" just covering the light bulb—look briefly to find the position. Stand back and switch on the "Sun." The "Moon" casts a shadow on the screen ("Earth").

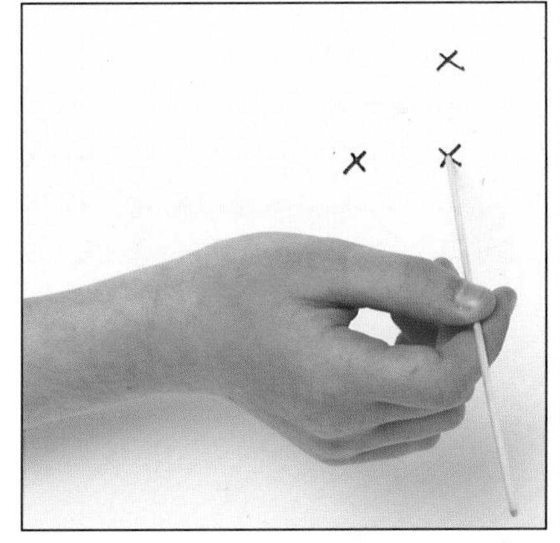

3 MARK THE SHADOW'S CENTER and two points at its edge. Punch holes at these points, and look briefly through each hole in turn to see the "eclipse."

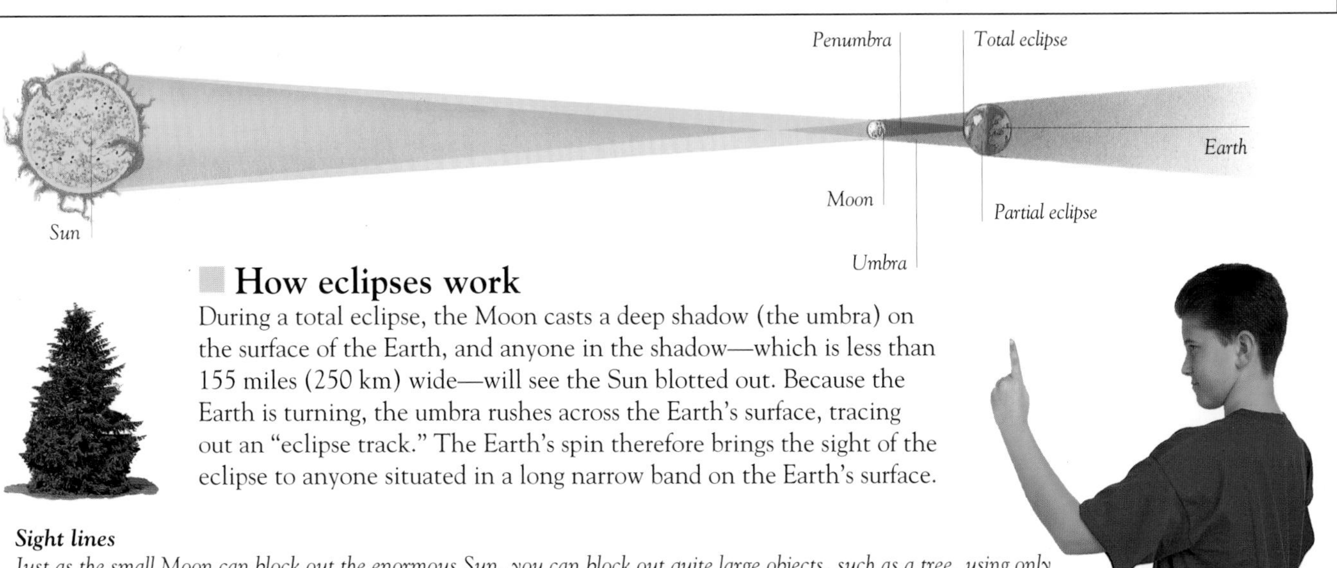

How eclipses work

During a total eclipse, the Moon casts a deep shadow (the umbra) on the surface of the Earth, and anyone in the shadow—which is less than 155 miles (250 km) wide—will see the Sun blotted out. Because the Earth is turning, the umbra rushes across the Earth's surface, tracing out an "eclipse track." The Earth's spin therefore brings the sight of the eclipse to anyone situated in a long narrow band on the Earth's surface.

Sight lines
Just as the small Moon can block out the enormous Sun, you can block out quite large objects, such as a tree, using only your finger—providing you position it correctly. Closing one eye makes it easier to see the object "disappear" completely.

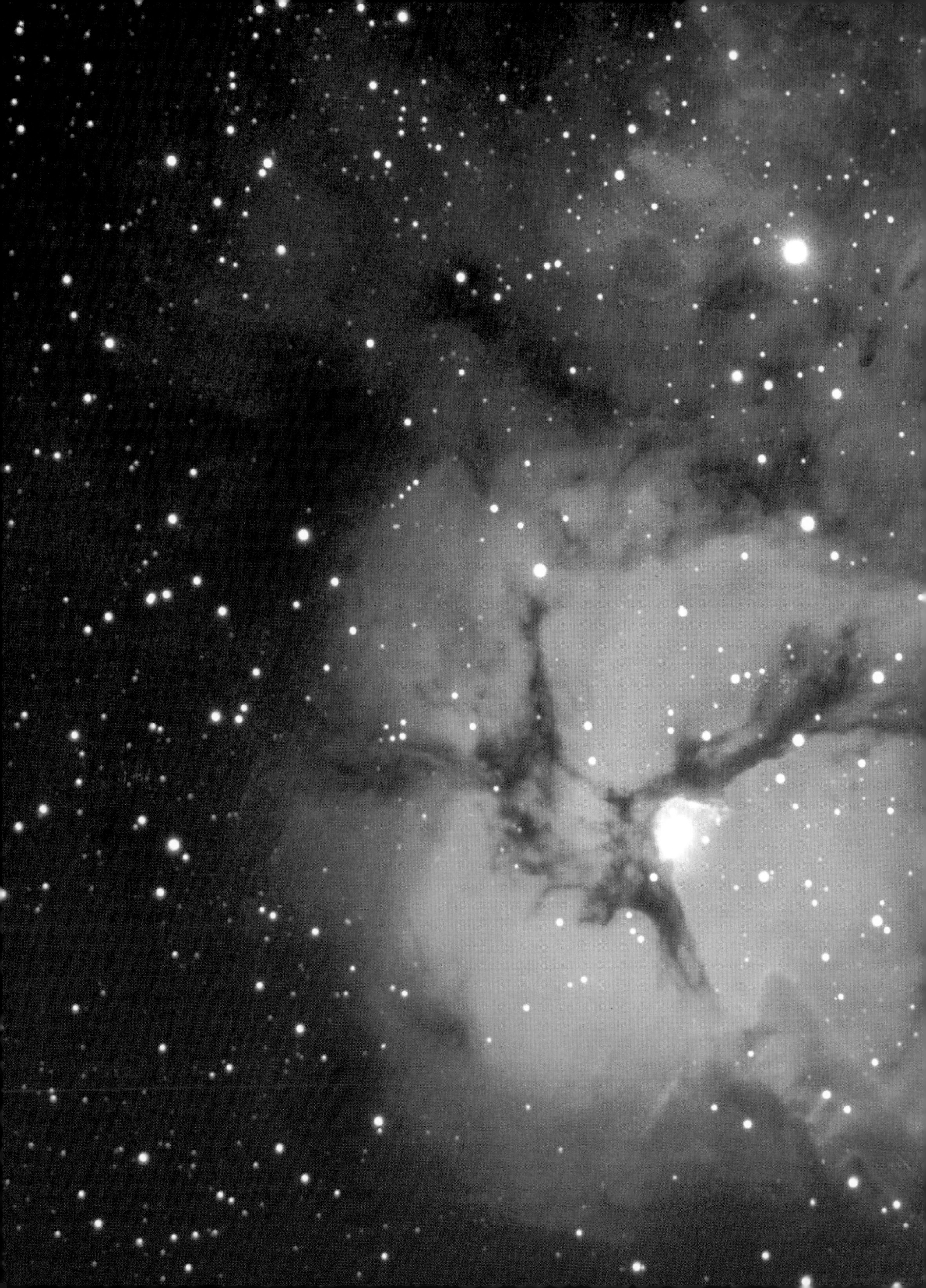

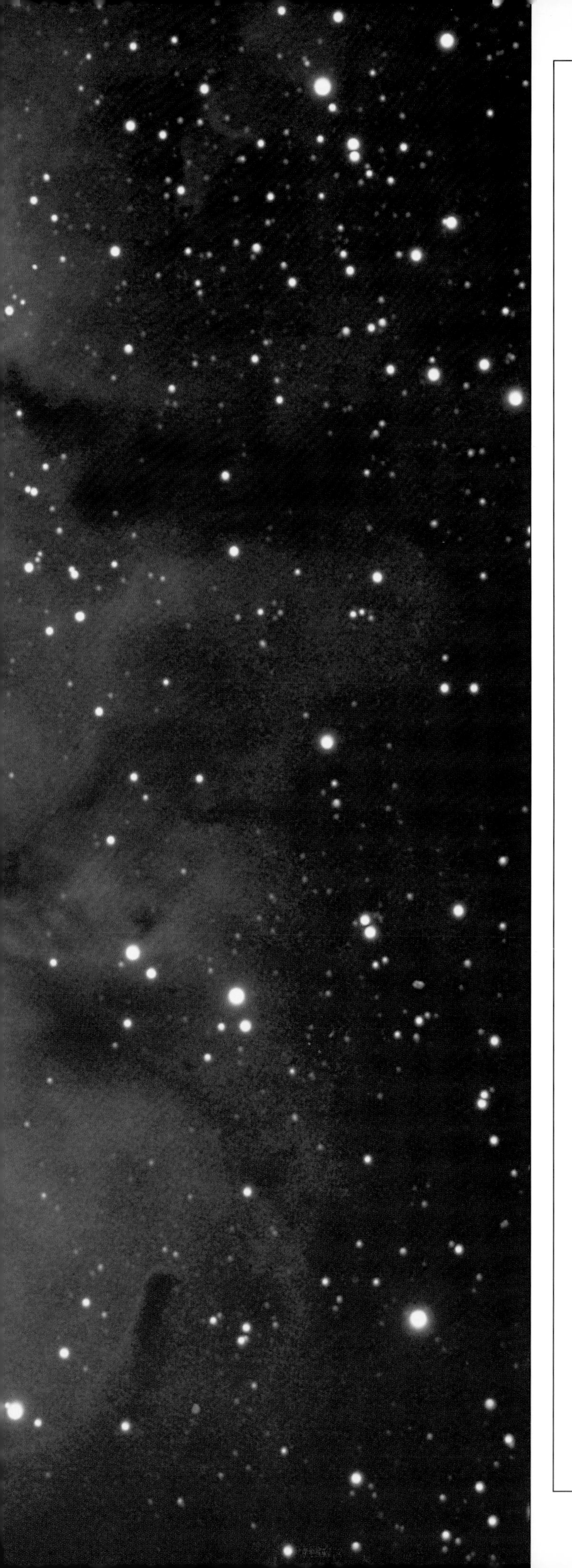

The STARS

Star birth
Stars are born in vast clouds of gas called nebulae. The Orion Nebula (part of which is seen above) is easily visible to the unaided eye. The Trifid Nebula (left) is divided by dark bands of dust—the raw material of planets.

OF THE 200 BILLION STARS IN our Galaxy, the Sun is a very ordinary star in terms of age, size, and brightness. Many of the other stars in our Galaxy are very different from the Sun—brighter, hotter, and bigger. Although even the nearest stars are almost 300,000 times farther away from us than the Sun, we can discover an astonishing amount about them—including how they are born, live, and die.

STARLIGHT AND STAR LIFE

ON A CLEAR NIGHT the sky is spangled with countless stars, looking like tiny diamonds against the black night sky. Each star is a blazing nuclear inferno like the Sun, but it is seemingly dimmed to a gentle glimmer by its immense distance. Our ancestors were not very interested in the physical nature of the stars. They used the stars in practical ways, to tell the time at night and the season of the year, to work out the direction of north, and to find their latitude on the Earth.

In order to use the stars, the early astronomers had to be able to tell which star was which. At first glance, this seems almost impossible. There seem to be millions of stars in the sky, so how do you even begin? In fact, the human eye can pick out only 6,000 stars in the entire sky—there are some 3,000 visible above the horizon at any one time. Even a slight trace of stray light, from the Moon or from streetlights, will drown out half of these. The only way to remember even the brightest stars, in practice, is to spot some of the patterns that they make. It is like a puzzle where you draw an outline by connecting certain dots—except in the sky you can select your own dots.

The Pleiades cluster
This beautiful star cluster lies in the constellation Taurus, about 410 light-years away. It is only 70 million years old—very young on the cosmic timescale. The bright blue-white stars are the heaviest and hottest in the cluster. They are lighting up a surrounding cloud of gas and dust.

Observing the stars
People have always been intrigued by the stars. Here an illuminated manuscript shows an Italian astronomer observing the stars through an early telescope.

Star patterns

All around the world, different peoples have made up their own star patterns—or constellations. And there is little agreement between them. Where the Greeks, for example, saw a queen sitting in a chair (Cassiopeia), the Chinese saw a chariot pulled by horses and a path across the mountains. But astronomers around the world now use the same constellations. They were first devised by the Babylonians around 5,000 years ago, and then passed on by the Greeks, the Romans, and the Arabs. Many are named after animals, such as Aries (the Ram) or mythological heroes (such as Perseus and Hercules). In the 17th century, European explorers named extra constellations in the very southern parts of the sky, which were not visible from the Middle East and the Mediterranean. These include some odd names, such as Caelum (the Chisel). At that time, Latin was the international language of science, and astronomers still use Latin names for the constellations. So the proper name for the Great Bear is Ursa Major, and the Southern Cross is officially called Crux. But most of the names of individual stars have come to us from Arab lands, which were the center of scientific learning around the year A.D. 1000. So Aldebaran (in Taurus, the Bull) is Arabic for "the follower," because it seems to chase the Pleiades (or Seven Sisters) across the sky. And Deneb, meaning "the tail," indeed marks the hind part of Cygnus, the Swan.

Although star patterns are essential for finding our way in the sky, the stars in each constellation are not really related to one another. They all lie at different distances from us. Some are nearby and are rather dim, while others are incredibly distant and enormously powerful

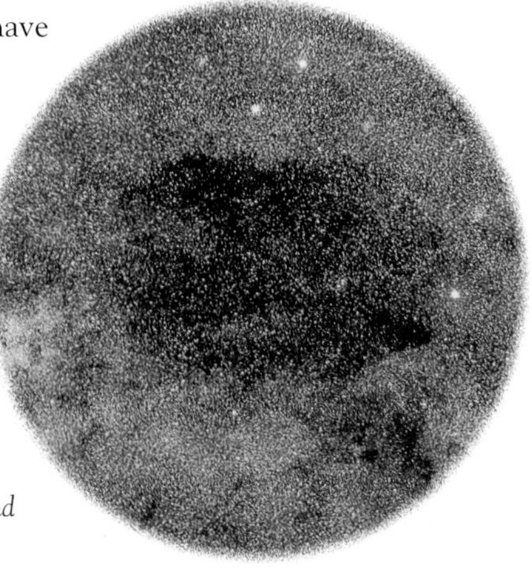

The Coal Sack
A dark cloud, the Coal Sack, is visible in the southern skies next to Crux, the Southern Cross. It is dark because it is thick with tiny grains of interstellar dust, which scatter and absorb the light from the stars beyond it. Part of Crux is visible at the upper right-hand side of this picture.

stars. The two main stars of Centaurus (the Centaur), for example, consist of a star very near to the Sun and another that lies 100 times farther away and is intrinsically thousands of times more luminous.

Light-years

Even the nearest stars to the Earth are at such vast distances that we can barely comprehend them. The closest star to the Sun, a dim red dwarf called Proxima Centauri, lies 25 million million miles away from us. The light from Proxima takes 4.2 years to reach us, so astronomers say it lies 4.2 light-years away. Deneb—the most distant from the Earth of the really bright stars—is 1,800 light-years away. If you traveled in a car, it would take 48 million years to reach Proxima Centauri. The journey would take the best part of 80,000 years for even the fastest space probes. In future centuries, we may build spaceships that can travel much faster. But the theory of relativity says that nothing can move faster than light, so even the speediest spacecraft would take a human lifetime to reach Aldebaran, 68 light-years away. Only with some unknown physics could people travel to farther stars.

Antares
This brilliant red giant—marking the heart of Scorpius (the Scorpion)—is a huge star near the end of its life.

Vital statistics

Once astronomers know the distance to a star, they can work out what the star is really like. The spectrum of a star's light contains information on its temperature and composition. In this way, astronomers have found that stars come in a bewildering variety of types. Some of them are hundreds of times bigger than our Sun: the red giant Betelgeuse, in the constellation Orion, would swallow up all the planets out to Mars if you swapped it for the Sun. On the other hand, a small companion to Sirius (the Dog Star) is a white dwarf (pp.118–119) that is no larger than our planet Earth. Spica, in Virgo (the Virgin) is a blue-white star several times hotter than the Sun, while our neighbor Proxima Centauri is a red dwarf, only half as hot as the Sun and smoldering such a dim, dull red that you need a telescope to see it at all.

One of the great scientific achievements of the 20th century has been to work out why the stars are so diverse. Much of the difference involves the fact that stars change as they get older.

A star's life

A star lives for so long—millions or even billions of years—that we cannot hope to see a star change from, say, a red giant to a white dwarf. But astronomers can deduce that these changes must happen. It is like an alien landing on Earth for a day: by studying babies, children, young adults, and old people the alien could work out how humans change with age, even though no one grew up or aged noticeably during that one day. In the case of stars, astronomers can see young stars still surrounded by the gas they were born from, middle-aged stars shining steadily, and large unstable stars that must have lived for so long that nuclear reactions have changed the composition of their centers.

A star's life begins in a dark cloud in space. The gas is pulled together by its own gravity to form a dense, glowing clump of matter. A star is born. For a long time, it shines steadily without changing. Eventually, the star expands and becomes a red giant. This star sheds its outer layers into space to leave a tiny core that fades from sight.

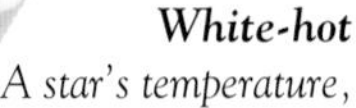

White-hot
A star's temperature, like that of a hot, glowing metal rod, can be determined by the color of the light that it emits.

Black hole
A white-hot disk of gas (foreground) hides a voracious black hole (p.152) in this artist's impression. The gravity of the black hole is pulling gas from the atmosphere of its companion, a blue giant star (background). As the gas spirals toward the black hole, it forms a flat "accretion disk," which heats up and emits X-rays.

Star theater

YOU CAN SEE HOW THE STARS look in the night sky—at any time of the day or night—with a star theater. The star theater will show you all the stars that are visible in either the Northern or Southern Hemisphere at any time of the year, so you don't have to wait for months to see them. You can also modify it to see how the stars appear to people on the opposite side of the world from you. You may have visited a planetarium, much larger kind of star theater. A planetarium has a complex projector that shows the positions of the planets and the Moon and how they move against their starry background. Simple star theaters are used to train navigators to find their way by observing the stars.

EXPERIMENT

Making a star theater

When making your star theater, you need to use five of the six squares shown here. Once you have identified the stars visible from your home, you can reassemble the theater by taking off the top square, turning the theater upside down, and putting the square you haven't used yet on as the lid. Now you can see how the sky looks from the opposite hemisphere—where it includes some stars that are never visible from your home.

YOU WILL NEED

● *poster board* ● *ruler*
● *pen* ● *scissors* ● *tape*
● *compass* ● *wooden skewer* ● *plastic putty*
● *4.5-V battery*
● *flashlight bulb and wire connections*
● *screwdriver*

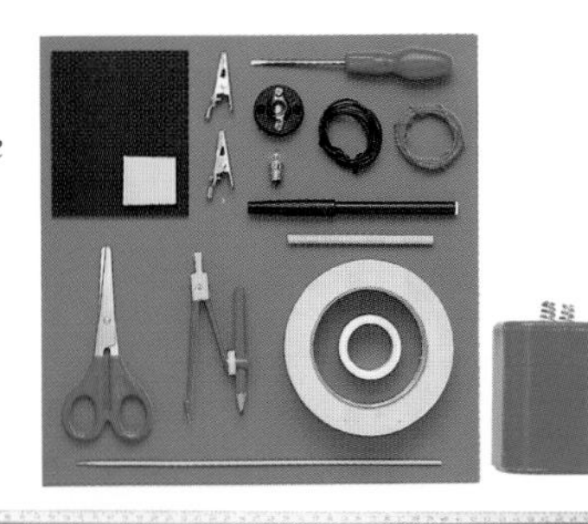

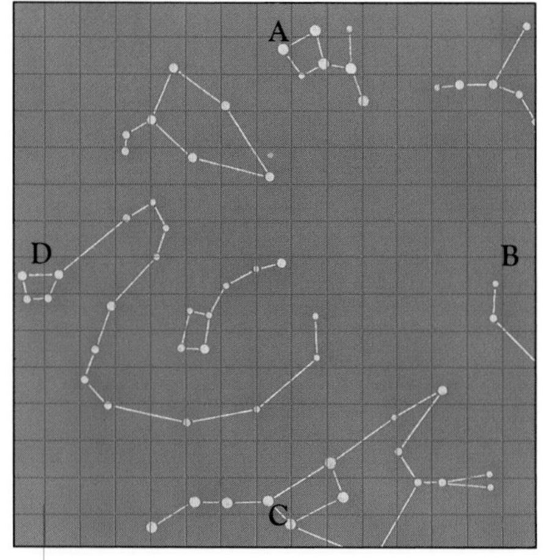

Roof of theater for Northern Hemisphere

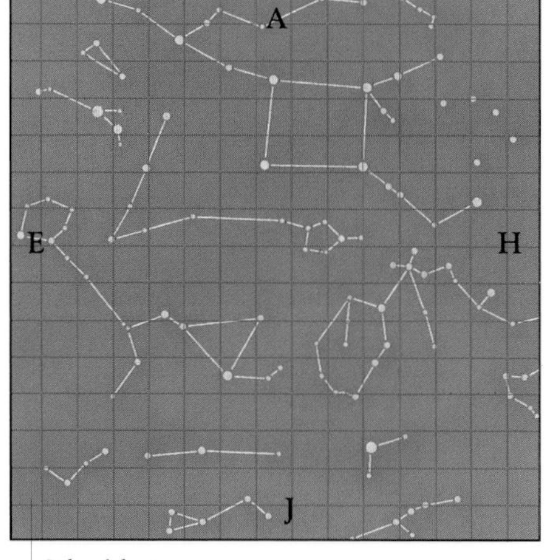

Side of theater

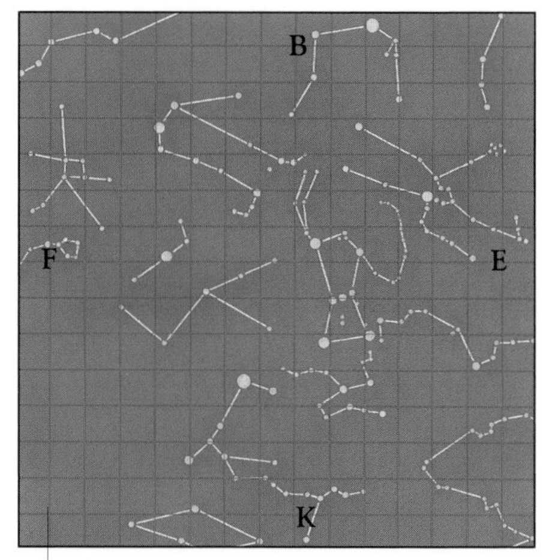

Side of theater

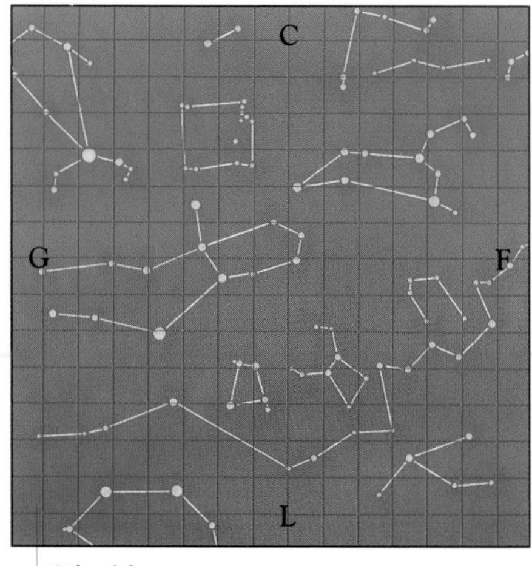

Side of theater

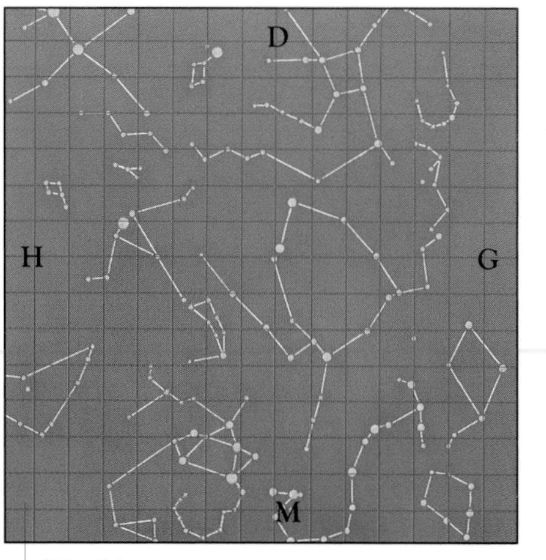

Side of theater

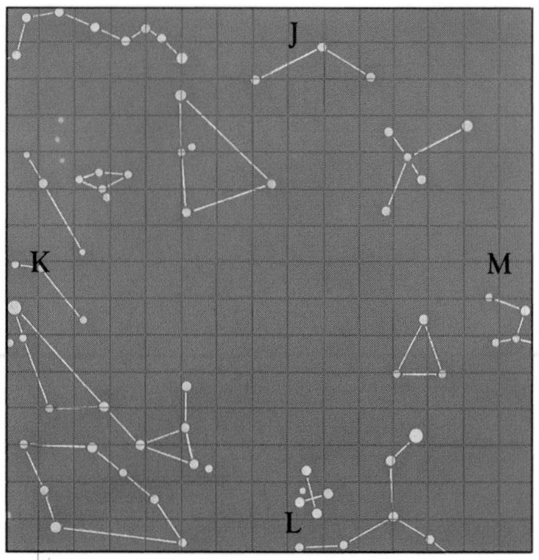

Roof of theater for Southern Hemisphere

1 TAKE SIX 15-in (30-cm) square pieces of poster board, and trace a 15 x 15 grid on each, making 1-in (2-cm) squares each way, so that you end up with 15 squares up each side, like the star grid plans shown opposite.

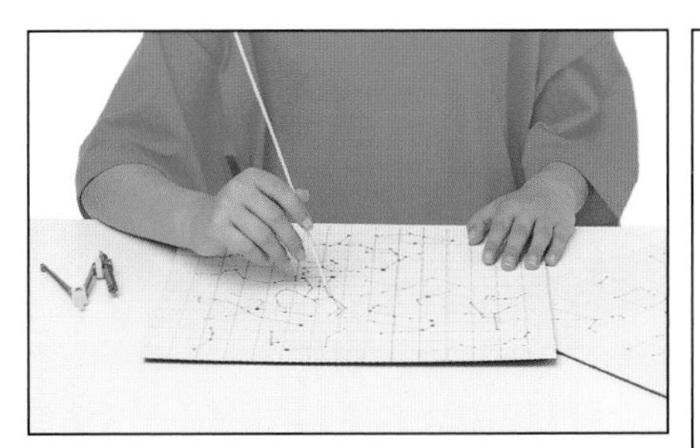

2 USE THE GRIDS to copy the constellations as shown on the plans. Then use the compass point to make large holes for the big stars and small ones for the smaller stars. Copy the letters to make assembly easier.

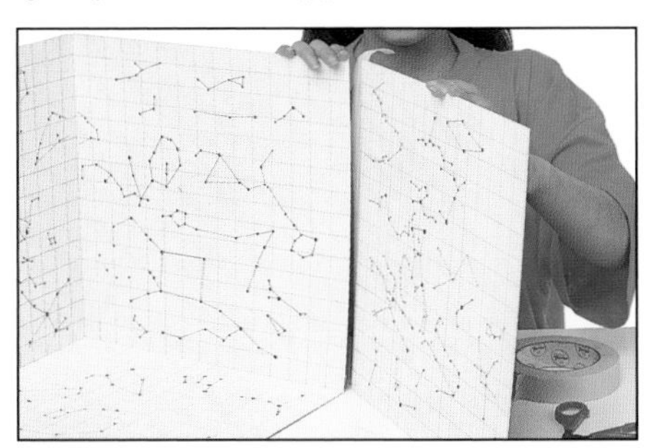

3 ASSEMBLE THE FIVE SQUARES for your hemisphere to form a box, with all the sides you drew the stars on facing inside. Match the letters at the edges of the plans opposite to figure out which edge to join to which. Fasten with tape.

4 MAKE A CIRCULAR BASE for the skewer out of poster board, and stick the two together with plastic putty. Mount the flashlight bulb in its holder on the skewer, and connect the wires to the bulb and to a 4.5-V battery.

The planetarium

Early planetariums were designed to show the public how the planets move, such as Perini's Planetarium (below) built in 1880. A Cambridge professor, Roger Long, constructed a large, hollow globe punched with holes. People sat inside the globe in the dark, and the holes shone brightly like stars. A German company, Karl Zeiss, built the first modern planetarium in 1923. It made stars by throwing light onto the inside of the dome from a central projector.

5 FIND A CLEAR TABLETOP or work surface for your theater. Turn on the flashlight bulb, and place the box over it. Switch off the lights to see the "stars" all over the room. Let your eyes adjust for a minute or two for the best results. You may have to turn the box slightly to make the "stars" really sharp. To identify the constellations, use the star maps on pages 148–151.

The constellations

PEOPLE LIKE to imagine the stars as connecting together into patterns or pictures. It is similar to linking one dot to the next in a puzzle, except that there is no right or wrong way to join up the stars. Different civilizations have picked out different patterns in the sky. Most of the star patterns—or constellations—that we use today were invented by the Babylonians and passed on by the ancient Greeks and Romans (which is why astronomers call constellations by their Latin names).

EXPERIMENT

A model constellation

Try making this scale model of a constellation with the stars at their relative distances, then look at it from different angles: the familiar shape we see from the Earth vanishes. The constellation here is Leo, visible from both hemispheres. You will have to follow the measurements below very carefully.

YOU WILL NEED
- *black poster board*
- *colored modeling clay*
- *black paint* ● *saucer*
- *paintbrush* ● *compass*
- *wooden skewers*
- *ruler with millimeters*
- *white crayon*
- *scissors*

Measurement chart (in millimeters)

Star	D	H	x	y
Denebola	10	105	00	10
Zosma	10	150	85	13
Algieba	14	165	255	25
Adhafera	12	225	270	30
Mu	14	240	345	45
Epsilon	16	225	375	77
Eta	20	135	300	450
Regulus	15	75	300	21
Coxa	10	96	85	19

1 MAKE NINE BALLS OF modeling clay with diameters (D) as shown in the chart above to represent the stars of Leo (roughly) and their real brightness.

2 MAKE CIRCULAR poster-board bases for the skewers, and paint the bases and the skewers black. For each star, cut a skewer to the height (H) in the chart.

3 ON A SHEET OF black poster board, use a white crayon to draw a grid of 10-mm squares, 380 mm wide and 450 mm long.

EXPERIMENT

Making constellations in a shoe box

You and your friends can test how well you recognize star patterns with these pierced "constellation cards." When held up to the light, the star patterns look surprisingly realistic.

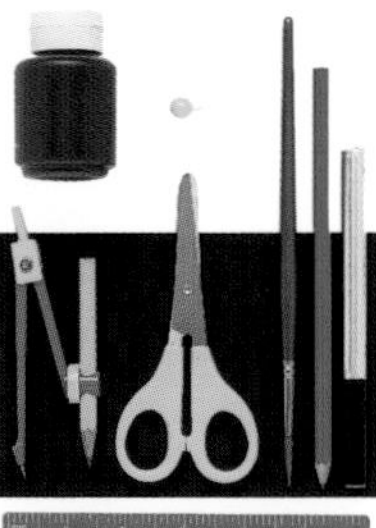

YOU WILL NEED
- *shoe box with lid* ● *black paint*
- *wide paintbrush* ● *ruler* ● *stiff black paper* ● *gold or white marker pen* ● *pushpin* ● *compass* ● *scissors*

1 CHECK THAT THE LID fits well to create a completely lightproof box. Using a wide brush, paint the shoe box and lid so that they are completely black on the inside.

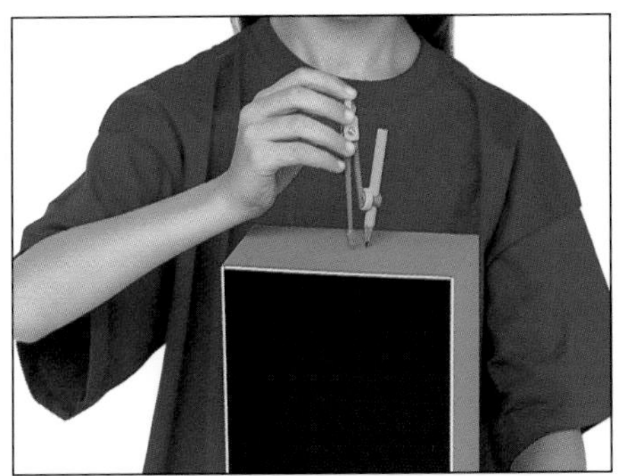

2 CUT A CIRCULAR HOLE about ½ in (1 cm) wide at one end. Then cut out the other end of the box, leaving 1 in (3 cm) on each side and at the base.

3 NOW CUT OUT rectangles from sturdy black paper, the same height but slightly wider than the rectangle you have cut out of the end of the box.

4 CHOOSE CONSTELLATIONS from the star maps on pages 148–151. With your bright marker pen, carefully mark the star positions on the rectangles.

5 WITH A PIN, make holes in the rectangles where you have marked stars. For brighter stars, make bigger holes.

6 ASK A FRIEND to put each rectangle in turn into the end of the box. Look through the viewing hole toward light. Do you recognize the constellation?

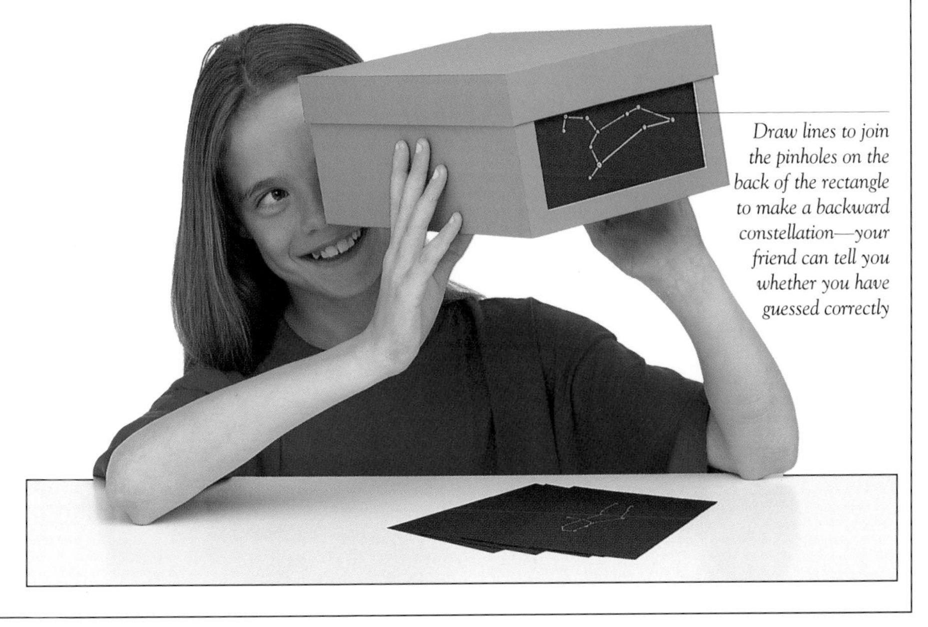

Draw lines to join the pinholes on the back of the rectangle to make a backward constellation—your friend can tell you whether you have guessed correctly

4 PUT THE "STARS" on the skewers. Position each "star": starting at the bottom left hand corner, go "x" mm across and "y" mm up to find its place.

5 THE VIEW FROM halfway along the bottom is the view of Leo seen from the Earth. Look from another angle to see how Leo's stars appear from other parts of space. Do they look like Leo?

Ask a friend to hold up stiff black paper to make a "space" background

Dialing the stars

THERE ARE STARS all around us in space, but at any one time, from any given place on Earth, only half of them are visible. The other half can be seen only from the opposite side of the Earth. The stars visible on any night depend on several things—your latitude (p.18) on the Earth, the time of year, and the time of night—so it is often difficult to know what you will see. But with a planisphere you can simply "dial up" the view of the sky for any date or time of night. A planisphere is a map of the sky. It consists of a circular star map for either the Northern or Southern Hemisphere (depending on where you live) and a rotating mask that reveals the stars at any time and date at your particular latitude. A planisphere cannot show where the planets are—they move around the sky too quickly to be marked.

EXPERIMENT

Making a planisphere

You can buy planispheres in stores, but a homemade one works just as well. The hole in the mask shows how the sky would look if you were lying on your back and looked up, with the edge of the hole being your horizon all the way around. To get the map the right way up, look north and turn the whole dial (without moving the mask) until the edge marked "north" is at the bottom. To look in other directions, turn the whole dial until the appropriate compass point on the mask is at the bottom. Where else on Earth would your planisphere work?

YOU WILL NEED

- *poster board* ● *glue*
- *tracing paper* ● *pen*
- *ruler* ● *compass*
- *paper fastener*
- *scissors* ● *acetate*

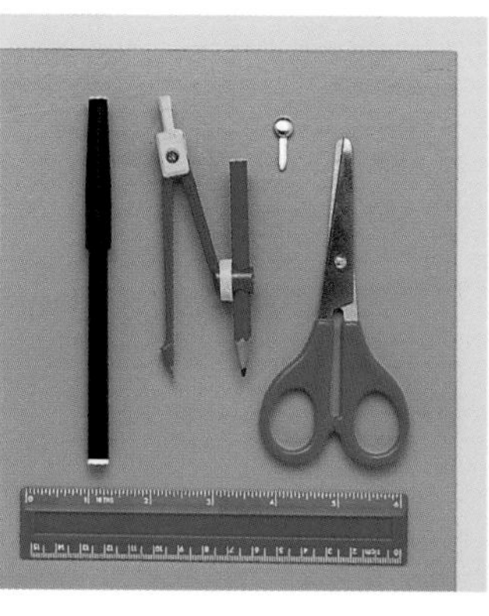

Making your dial
Copy the inner values and times for the Northern Hemisphere and outer values for the Southern Hemisphere. Note that the compass points will be reversed from the usual map direction.

Divide each month into four quarters, or weeks, so you can dial the date more easily

Cutout for 35° (southern Europe, southern U.S., Australia, South Africa)

Cutout for 42° (northern Europe, northern U.S., New Zealand)

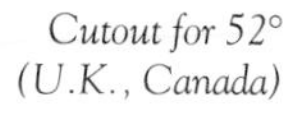

Cutout for 52° (U.K., Canada)

Fall skies
This northern planisphere shows the sky for 42° north latitude, at 10 p.m. on September 20 (or 8 p.m. on October 20, and so on).

1 USE A COMPASS to trace a circle measuring 9¾ in (245 mm) in diameter on poster board. Cut out the circle—the base for the planisphere. Now trace a circle of the same size on tracing paper. Cut this out, and set it aside.

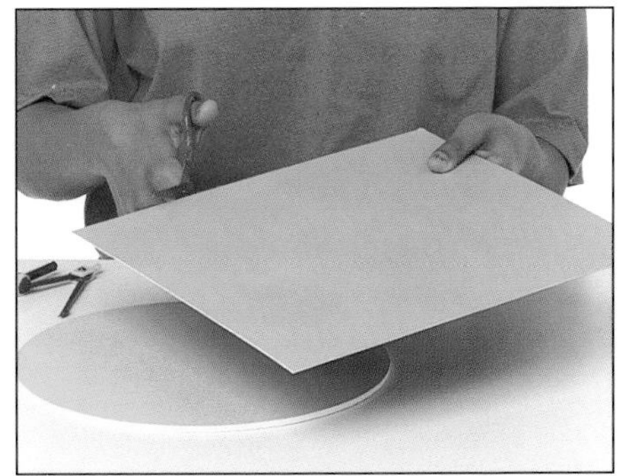

2 CUT OUT a poster-board mask measuring 9¼ in (235 mm) in diameter. Find your latitude (pp.18–19), and cut a hole in the mask for the latitude nearest yours. Mark the 24-hour clock and the compass points as shown opposite.

3 PLACE THE TRACING PAPER over the star chart for the appropriate hemisphere (pp.148–151), and copy the stars very carefully. Include the months around the edge, and draw six longitude lines across the diameter.

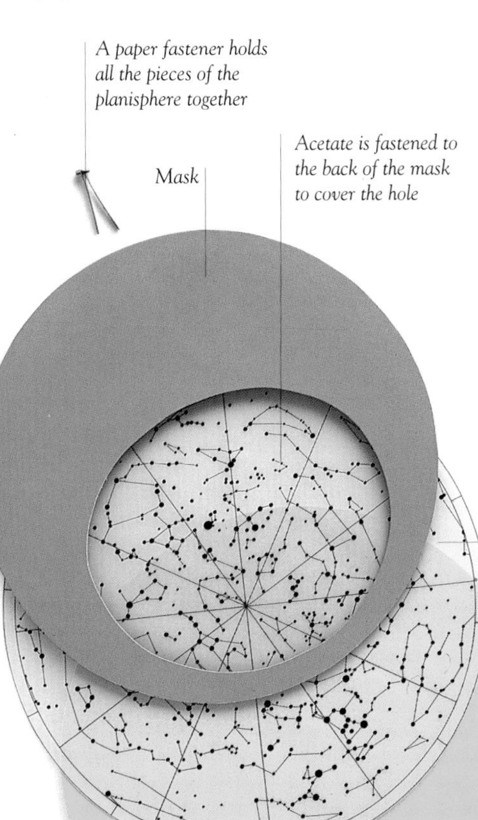

A paper fastener holds all the pieces of the planisphere together

Mask

Acetate is fastened to the back of the mask to cover the hole

Star map

Base

Look to the north (south if you live in the Southern Hemisphere), and turn the dial so that "north" (or "south") is at the bottom—this will show you the stars you should be able to see from where you live

4 CUT OUT A PIECE of clear acetate that is big enough to cover the hole in the mask. Glue this piece of acetate to the back of the mask. Once you have done that, glue the star chart to the base and leave it to dry.

5 PUNCH A hole in the center of the mask (through the acetate) and in the center of the star chart and the base, and insert a paper fastener to hold them together. Find today's date around the outer edge of the dial. Turn the mask until the right time lines up with the date. Hold the dial as shown, and look in the direction given at the base of the mask.

Starlight

SINCE THE STARS ARE so far away, the only way to learn about them is by examining the light that they produce. Looking at a starry sky, you will probably notice two things about the stars: they have different brightnesses, and they twinkle. "Twinkling" is caused by the Earth's atmosphere, but stars really do differ in brightness. If you can find out how far away a star is, you can measure how much energy it produces as light—its luminosity. Some stars, "variable stars," actually change slowly in luminosity. By studying a star's light with a spectroscope, astronomers can discover its temperature, composition, and quirks.

How bright are the stars?

The ancient Greek astronomer Hipparchus (writing in 146–127 B.C.) classified the most brilliant stars as "first magnitude," fainter ones as "second magnitude," and so on down to "sixth magnitude" for the very dimmest stars you can see with the naked eye. These diagrams show the magnitude of a bright star and a faint star in each of several well-known constellations. Spot them on a clear night, and estimate the magnitude of the other stars shown.

Gamma

Acrux (1st magnitude)

CRUX, OR SOUTHERN CROSS

Zeta (4th magnitude)

Crux

Crux, commonly called the Southern Cross, is the smallest constellation in the sky. Its compactness makes it a striking sight in the skies of the Southern Hemisphere. Of the four main stars of the cross, three are white or bluish white, but Gamma (top) is red.

Antares (1st magnitude)

Tau (3rd magnitude)

SCORPIUS

Scorpius

A constellation that really looks like its namesake, Scorpius is best seen from the Southern Hemisphere and the equator. The first-magnitude star Antares, which marks the scorpion's heart, is 300 times bigger than the Sun. It is deep red in color, and its name means "rival of Mars."

VIRGO

Theta (4th magnitude)

Spica (1st magnitude)

Virgo

In contrast to Crux, Virgo is the second-largest constellation in the sky. Its main stars make up a large Y-shape (right) with the first-magnitude star Spica marking the base of the Y. In legend, the constellation represents the goddess of the harvest, and Spica is pictured as an ear of wheat.

EXPERIMENT

Why the stars twinkle and the planets do not

Stars twinkle because of the turbulence of the Earth's atmosphere. Pockets of air at different densities are continually moving in front of the stars, making the light of stars appear to flicker. Planets, however, are so close (relatively speaking) that they appear to cover a bigger area than the individual air pockets, so their apparent brightness is relatively unaffected by the air's movement. Until recently, twinkling was a problem for astronomers monitoring stars through large telescopes, because twinkling also bends starlight in different directions. The newest telescopes have "adaptive optics"—computer-controlled systems that react to atmospheric turbulence. You can investigate the effects of air turbulence on "stars" and "planets" with a friend.

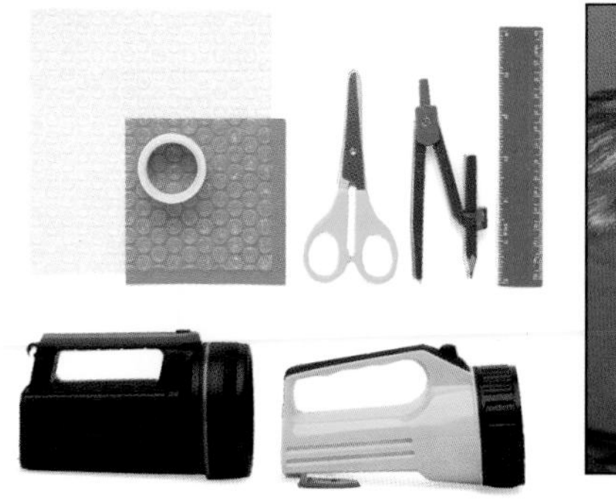

YOU WILL NEED

- *bubble wrap* • *paper* • *tape* • *2 flashlights*
- *scissors* • *ruler* • *compass*

Altair (1st magnitude)

Alshain (4th magnitude)

AQUILA

Aquila
The ancient constellation of Aquila (the Eagle) was seen in Roman mythology as the bird who carried Jupiter's thunderbolts. Its first-magnitude star, Altair, is one of the Sun's nearest neighbors, only 16 light-years away —93 million million miles.

Variable stars

Some stars brighten or fade over periods of days or months. Sometimes the star itself has a dimmer companion obscuring it: an "eclipsing binary" (p.121). Others, like red giants (pp.118–119), change in brightness as they swell and shrink. This graph is a "light curve" showing the changing brightness of the red giant Mira.

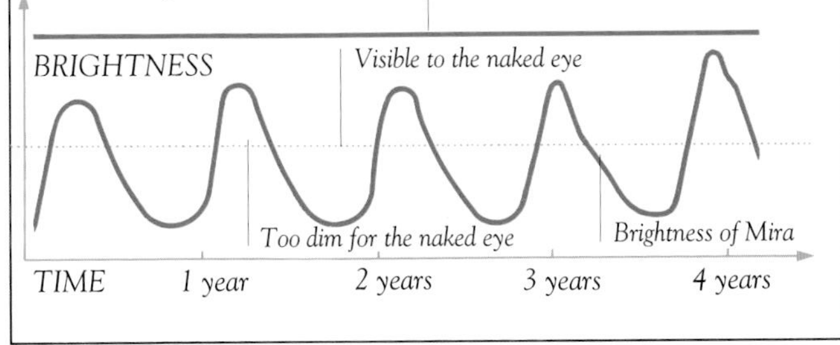

AURIGA

Capella (1st magnitude)

Pi (4th magnitude)

Auriga
Auriga (the Charioteer) includes among its stars first-magnitude Capella—the star that is overhead in midnorthern latitudes on winter evenings. Capella is a yellow star, similar in temperature to our Sun. However, it lies 42 light-years away and is actually 60 times brighter.

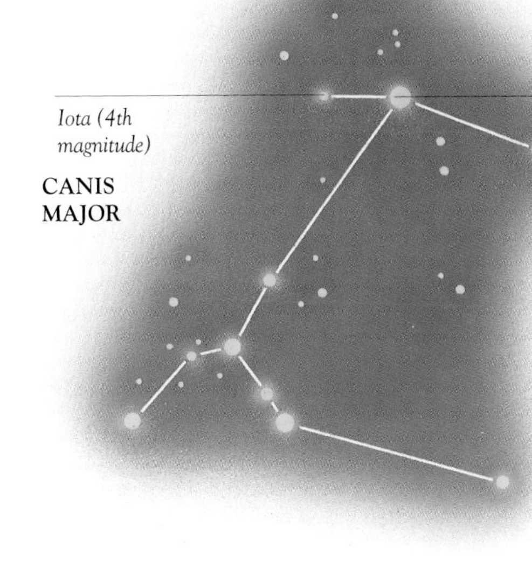

Sirius (1st magnitude)

Canis Major
Canis Major, the larger of the two "dogs" belonging to Orion, contains the brightest star in the sky—Sirius. Strictly speaking, Sirius is so brilliant that it has a negative magnitude (–1.46). In reality, Sirius is not that luminous. Like Altair, it happens to lie nearby: 8.6 light-years away.

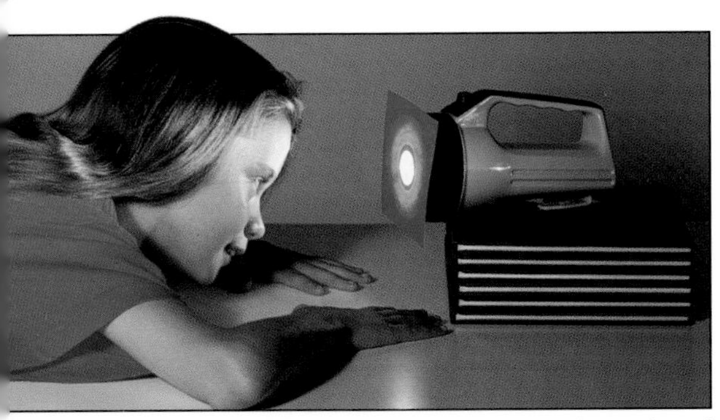

1 SET THE FLASHLIGHTS 16 ft (5 m) away ("star") and 1 ft (30 cm) away ("planet"). Cover the "planet" with a piece of paper with a $1\frac{1}{4}$-in (3-cm) hole cut out.

2 ASK A FRIEND to move the bubble wrap (the "atmosphere") back and forth in front of the "planet." Does its brightness vary? Now do the same for the "star." What difference do you see? If you go out at night, you can see the stars twinkling for real. Which stars twinkle most—stars that are overhead or those close to the horizon? Why?

How far are the stars?

AT FIRST GLANCE it may seem as if all the stars were at the same distance from the Earth, pinned to the dome of the sky. The nearest star to us after the Sun is Proxima Centauri—a faint companion to Alpha Centauri—which is 25 million million miles (40 million million km) away. These vast distances are measured in light-years, the distance light travels in a year—about 6 million million miles (9.5 million million km). Proxima Centauri is over 4 light-years away—its light left it 4 years ago.

■ DISCOVERY ■

Friedrich Bessel

Until the mid-1800's it was not known how to measure a star's distance with any accuracy. The German astronomer Friedrich Bessel (1784–1846) was a pioneer in this field and established the exact positions of about 50,000 stars. He was the first person to measure the distance to a star by parallax (see below). Using this method, he calculated the location of a star that is barely visible to the naked eye, 61 Cygni.

EXPERIMENT

Shifting stars

Some stars seem to shift their positions against the background of distant stars if they are viewed from different positions. You can try this "parallax" effect using homemade stars.

YOU WILL NEED
- *dark paper*
- *stick-on stars*
- *pencil* • *tape*

1 STICK SOME STARS onto a large sheet of dark paper measuring about 1½ x 2 ft (45 x 60 cm). This represents the distant stars.

2 ATTACH ANOTHER STAR to the side of the pencil. This represents a star that is near the Earth. Now tape the sheet of paper to a wall.

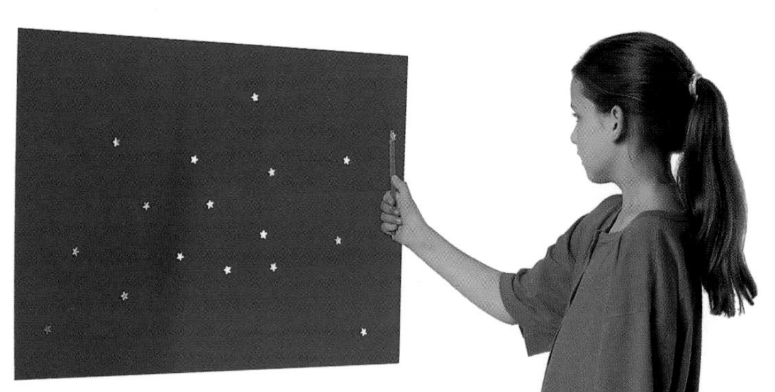

3 STAND 6 FT (180 cm) in front of the paper. Hold the pencil half an arm's length in front of you so that the pencil's star appears in front of the paper. Shut your left eye, and note the position of the pencil's star. Then switch eyes, and see where the star has moved. The difference in position is caused by parallax.

4 HOLDING THE PENCIL at arm's length repeat Step 3 to see the parallax change for a more distant star.

The principle of parallax

A star appears to shift its position against the background of more distant stars if it is viewed from opposite sides of the Earth's orbit. From the size of this "parallax shift" and the diameter of the Earth's orbit, astronomers can calculate a star's distance using simple geometry. The farther away the star is located, the smaller the parallax shift. In practice all shifts are minuscule—less than 1/3,000 the size of the Moon's disk as seen in our skies.

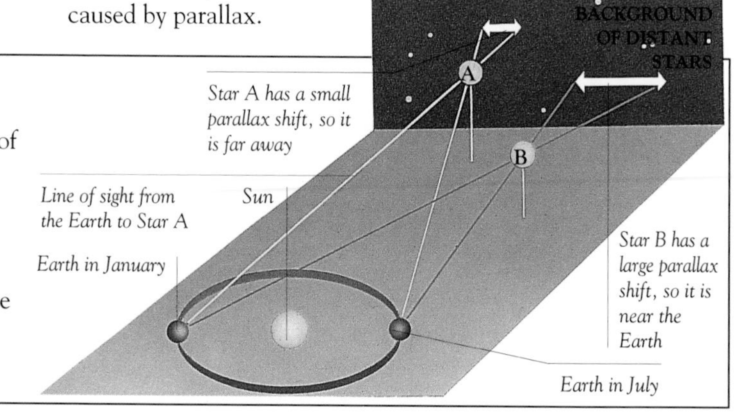

EXPERIMENT

The nearest stars

Because the distances between stars are so incredibly large, astronomers simplify them by measuring in light-years. Our closest star is the Sun. Its light takes just 8 minutes to reach us. The next closest star is Proxima Centauri, which is 4.2 light-years away. Understanding the distances between stars is easier using a model. Here you can plot the three-dimensional positions of our nearest stars.

YOU WILL NEED

● *large sheet of dark foamcore* ● *colored beads* ● *wooden skewers* ● *ruler* ● *pen* ● *scissors* ● *white paint* ● *paintbrush*

Southern stars

Four bright stars, called the Southern Cross, form a diagonal shape in the New Zealand sky. The Southern Cross is visible throughout the Southern Hemisphere. The two bright stars on the left of the picture are Alpha Centauri and Beta Centauri. Although they seem almost equal in brightness, Beta Centauri is about 100 times farther from the Earth than Alpha Centauri.

Horizontal positions
Use the grid above as a guide to plot the exact horizontal locations of the stars on the foamcore. Use the chart at right to give your "stars" the correct height. One light-year is 1/4 in (6 mm) in this scale. This shows the view of our nearest stars as seen from far away in space.

1 DRAW A 12 X 12 GRID of 1-in (25-mm) squares on foamcore. Paint white dots for the stars' positions (use the grid at left for reference).

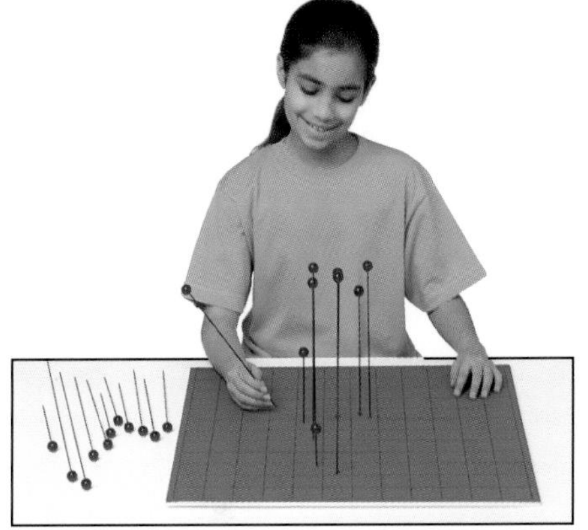

2 CUT 17 SKEWERS to the lengths shown below. Push a bead on one end of each skewer, using a different color for the "Sun." Place the skewers on the grid in the correct positions.

	STAR	SKEWER LENGTH
A	*Alpha Centauri*	*6¼ in (16 cm)*
B	*Altair*	*6¼ in (16 cm)*
C	*Arcturus*	*10½ in (27 cm)*
D	*Beta Hydri*	*2 in (5 cm)*
E	*Capella*	*18 in (45 cm)*
F	*Castor*	*17 in (43 cm)*
G	*Epsilon Eridani*	*8½ in (22 cm)*
H	*Eta Cassiopeiae*	*9 in (23 cm)*
I	*Fomalhaut*	*4 in (10 cm)*
J	*Mu Herculis*	*8¼ in (21 cm)*
K	*Muphrid*	*10¼ in (26 cm)*
L	*Pi-3 Orionis*	*7 in (18 cm)*
M	*Pollux*	*12¼ in (31 cm)*
N	*Porrima*	*10 in (25 cm)*
O	*Procyon*	*8¼ in (21 cm)*
P	*Sirius*	*8½ in (22 cm)*
Q	*Sun*	*7 in (18 cm)*
R	*Tau Ceti*	*7 in (18 cm)*
S	*Vega*	*9 in (23 cm)*
T	*Zeta Herculis*	*9 in (23 cm)*

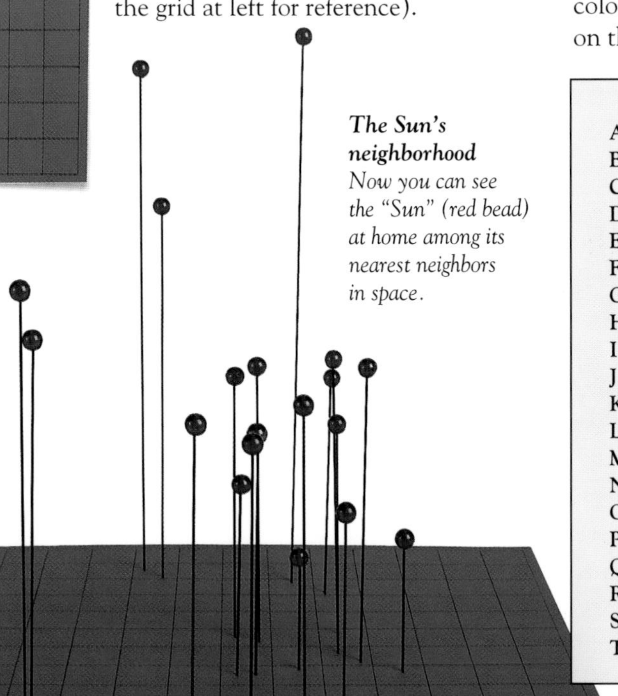

The Sun's neighborhood
Now you can see the "Sun" (red bead) at home among its nearest neighbors in space.

Red giants and white dwarfs

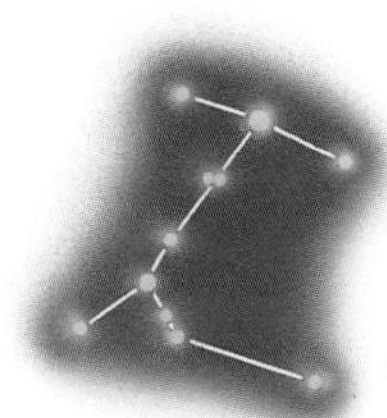

ALTHOUGH THE STARS may seem superficially similar, when you study them in depth you find they are a very mixed bunch. Some are much brighter than the Sun; some are much fainter. Stars that are much hotter than the Sun shine blue-white; others are cooler and red in color. And some stars are much bigger than the Sun, while other stars are relatively tiny. Two aptly named kinds of star illustrate the range. Red giants are over 100 times wider than the Sun and shine with a deep red glow. White dwarfs are much smaller than the Sun (they are no larger than the planet Earth) and have a searing surface that shines white-hot.

A star's life

Although stars differ in their brightness, mass, and temperature, they all share a similar life story (unless they are very massive—see pp.124–125). This is the story of a star like the Sun. The more massive stars live for a shorter time, while lightweight stars can live for up to 100 billion years.

White-hot

Many stars appear white because they are so hot—around 18,000° F (10,0000° C). But not all stars are white. In fact, a star's color is a guide to its temperature. Red stars are the coolest, and blue-white the hottest. When a substance—such as this iron bar—is heated, its color changes from dark gray to dull red. As it gets hotter, the color shifts from red, through orange, to yellow, to almost white. If it gets even hotter, it will glow pure white and then blue-white. At the white stage, the iron bar begins to melt. You can demonstrate this if you have a light with a dimmer switch. If it is just barely turned on, the light emits a faint red glow, then yellow as it is turned up, and finally white.

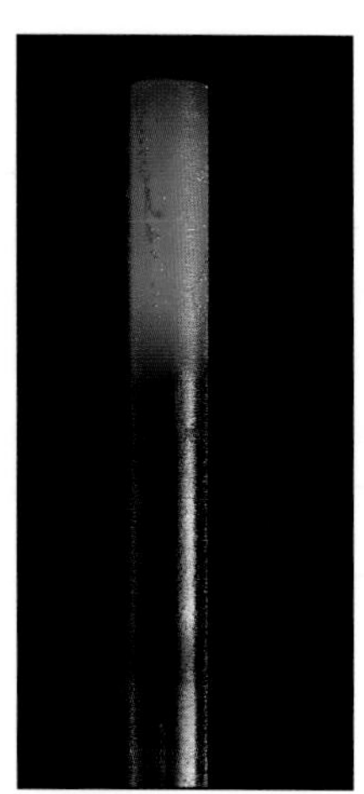

Red light
When the iron bar is heated, it begins to lose its original dark coloring and glow a dull red. The iron atoms emit light as the heat energy makes them vibrate.

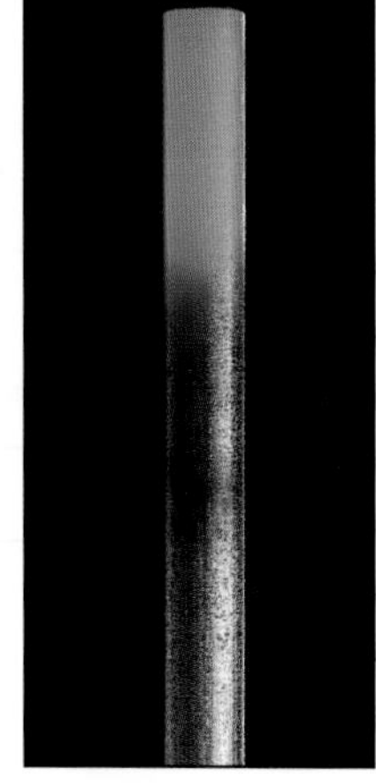

Orange light
As the bar continues to be heated, the iron atoms vibrate more quickly. They now emit an even brighter light, which has changed in color from red to orange.

Yellow light
The metal is now extremely hot, and the tip of the iron bar glows a bright yellow in color. However, the cooler parts of the bar still emit orange and red light.

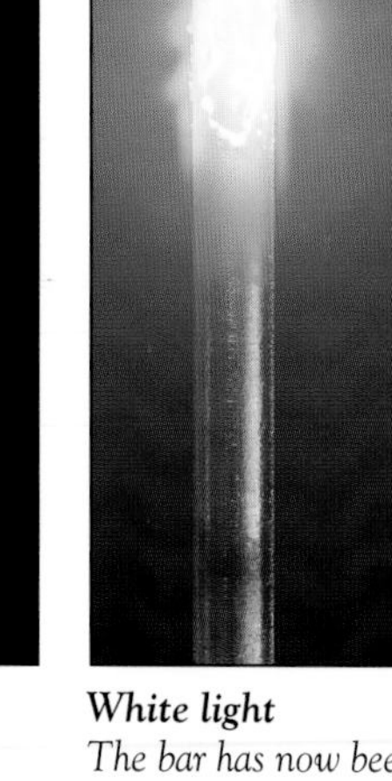

White light
The bar has now been heated to its melting point. The heat is so intense that the bar is white-hot. If it could survive melting, the bar would shine blue-white.

The temperatures of the stars
With the information below, go outside at night and take the temperatures of these stars: Betelgeuse, Capella, Rigel, Sirius, Canopus, Antares, Deneb, Fomalhaut, Acrux, Aldebaran, Achernar, Alpha Centauri, and Vega. Use the star charts (pp.148–151) to find them. Binoculars help show the colors.

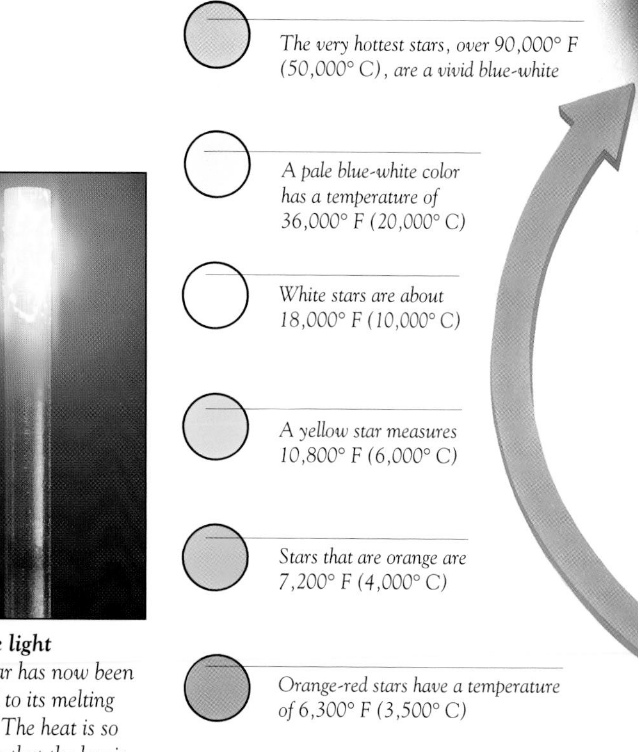

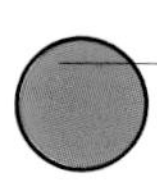

1 Beginning
A dense cloud of dust and gas starts to collapse under its own gravity.

2 A star is born
Individual clumps continue to collapse, growing hotter and hotter. When a clump's central temperature reaches 18 million° F (10 million° C), nuclear fusion reactions start, and a star is born.

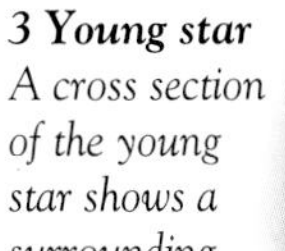

3 Young star
A cross section of the young star shows a surrounding cavity in the dark cloud blown by the star's fierce solar wind.

4 Forming planets
Gas and dust not incorporated into the young star may eventually form into a system of planets.

8 Planetary nebula
The star gently jettisons its unstable atmosphere as a planetary nebula, revealing its collapsed core.

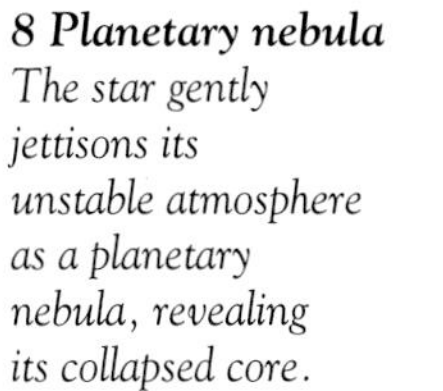

9 White dwarf
As the atmosphere drifts away, the collapsed core survives as an intensely hot white dwarf.

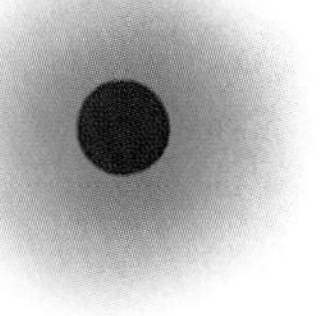

10 Star death
The white dwarf slowly cools to become a cold black globe.

5 Nuclear reactor
For billions of years the star shines by converting hydrogen to helium in its core.

6 Steady star
Even after 9 billion years, the star is virtually unchanged.

7 Red giant
After about 10 billion years, the central hydrogen runs out. The core collapses and heats up; the star's atmosphere expands and cools. The star becomes a red giant.

Double trouble

VERY FEW STARS are single; most have at least one companion. Pairs of stars, called binaries, are usually so close together that you cannot see them as separate with the naked eye. In some cases, four or six stars live together: one pair orbits around another star or another pair of stars, with other pairs going around them. The Earth lies directly in line with the orbits of several double stars. To us the two stars appear to pass in front of and behind each other as they orbit, regularly eclipsing one another. These eclipses cause the light of the system to vary, particularly—as in the case of Algol in Perseus—if one star is much brighter than the other and is sometimes covered by the dimmer star. Because the two members of a binary look like a single star from Earth, the eclipses give the appearance of a star regularly varying in brightness.

EXPERIMENT

Build a star mobile

This is a scale model of a star system called Almach, which forms part of the constellation Andromeda. From a distance, it looks like a double star. But one of these "stars" is triple, consisting of a close pair of stars orbiting a third star. The stars have different temperatures, which make them shine different colors. When the model is complete, try to imagine life on a planet going around one of these stars. Would you ever have nighttime? With these four "Suns" how many shadows would you have? What color would your shadows be? You can also invent your own star mobile, complete with blue stars, red giants, white dwarfs, and so on. To make it look real, build up from pairs of stars.

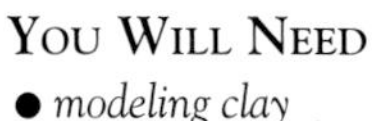

YOU WILL NEED
- *modeling clay*
- *string* • *scissors*
- *wooden skewers*
- *ruler*

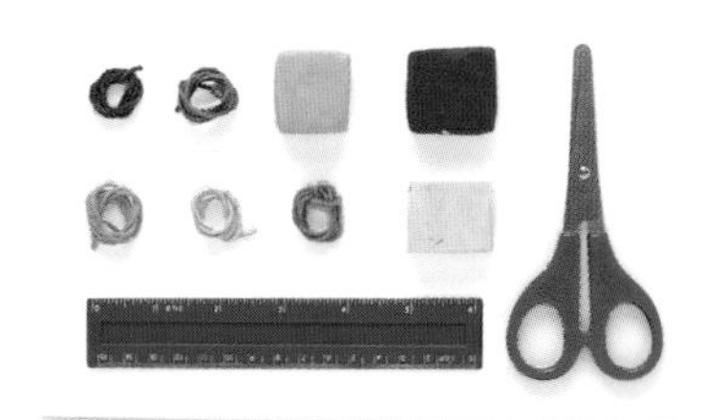

1 CAREFULLY CUT five pieces of string to measure 12 in, $4\frac{3}{4}$ in, $1\frac{1}{2}$ in, $\frac{5}{8}$ in, and $\frac{3}{4}$ in. Then cut three lengths of skewer to measure $\frac{1}{2}$ in, 1 in, and $8\frac{1}{4}$ in.

2 MAKE FOUR BLOBS of clay: two blue and one white (each $\frac{3}{8}$ in wide), and one yellow ($\frac{3}{4}$ in wide). Put the blue blobs at each end of the $\frac{1}{2}$-in skewer.

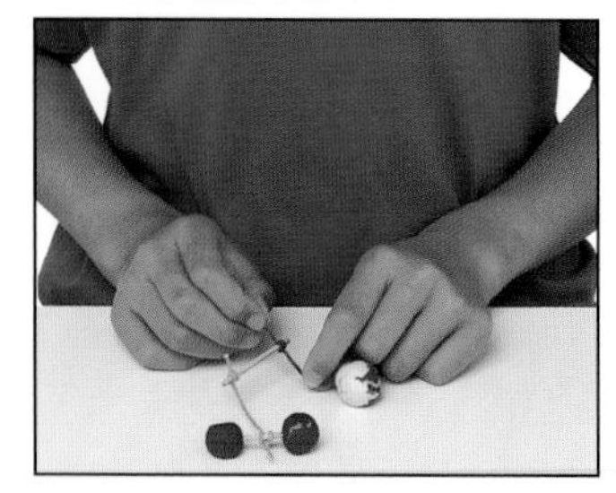

3 TIE ON THE $\frac{3}{4}$-in string. Tie the free end to one end of the 1-in skewer. To the other end of the 1-in skewer, attach the white blob and the $\frac{5}{8}$-in string.

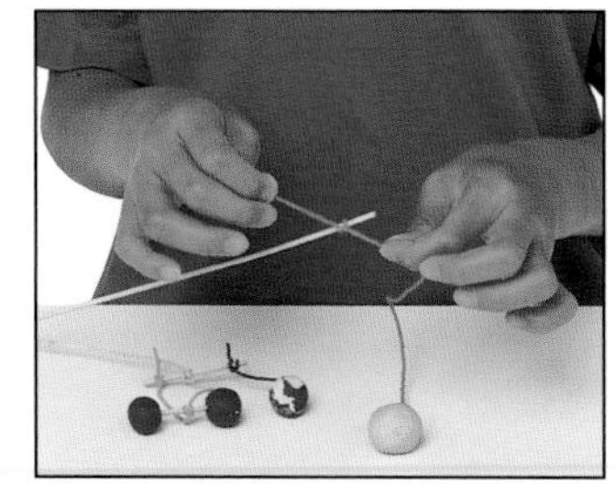

4 TIE THE $1\frac{1}{2}$-in string to the middle of the 1-in skewer. Tie the free end to one end of the $8\frac{1}{4}$-in skewer. To the other end, tie the yellow blob, with the $4\frac{3}{4}$-in string.

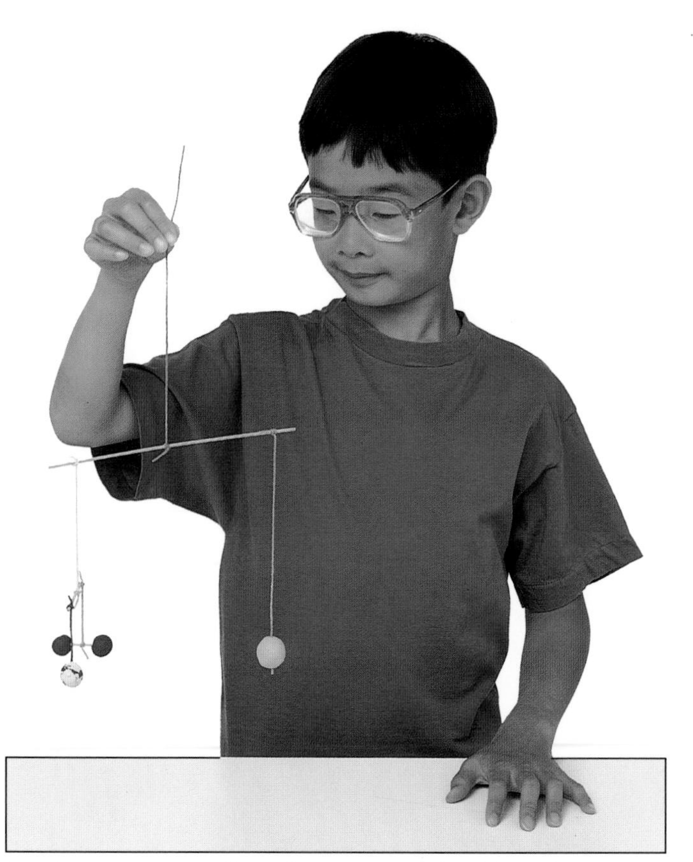

5 FINALLY, TIE THE 12-in-long string to the $8\frac{1}{4}$-in skewer. Hold up your mobile, and make the star system balance by moving the string left or right along the skewer.

Finding double stars

Some double stars cannot be seen as separate even with a telescope, and astronomers must rely on indirect clues, such as one star moving in front of its companion and hiding its light. But there are some interesting double stars—or stars that appear double—that can be seen with the naked eye or binoculars. Use these diagrams to find them, locating the constellations using star maps (pp.148–151) and your planisphere (pp.112–113).

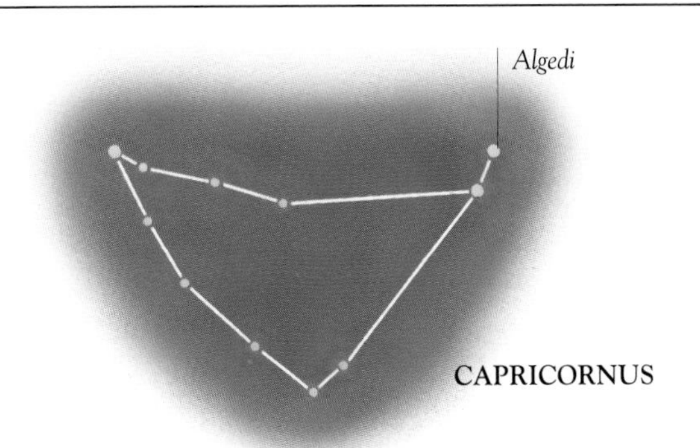

Algedi
This unrelated pair of stars is visible to the unaided eye. Through a telescope, one of these stars looks double, while the other is a triple star system.

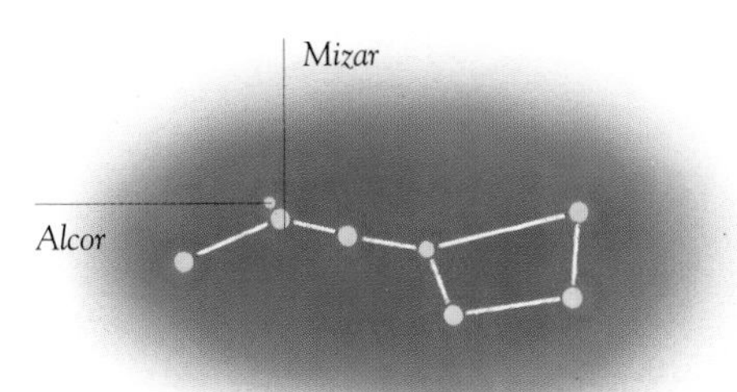

Mizar and Alcor
These two stars form a very obvious double to the unaided eye, but they are not actually associated. Alcor lies farther away than Mizar, and just happens to appear almost behind it. Through a telescope, Mizar is double.

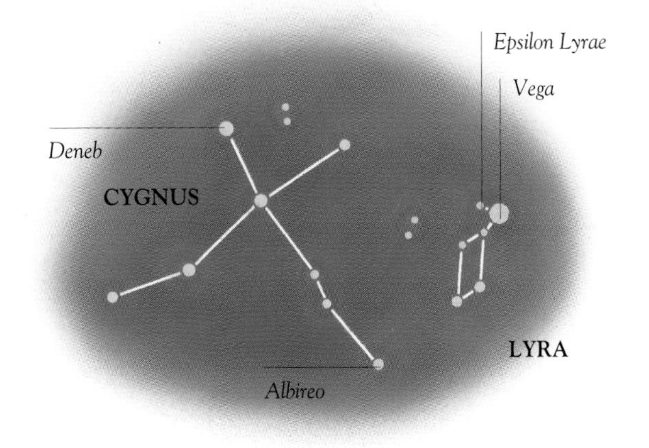

Albireo and Epsilon Lyrae
In a telescope Albireo is one of the most beautiful doubles —a bright yellow star circled by a fainter blue companion. Epsilon Lyrae is visible as double to the unaided eye or with binoculars; with a telescope, each star is also a double.

Alpha Centauri
The nearest star system to the Sun is revealed as a pair through a small telescope. Proxima Centauri (too faint to be shown here), an outlying member of the system, is the nearest star to the Sun.

EXPERIMENT

An eclipsing binary

You can simulate a binary star's motion using a small bright ball as the more brilliant star and a large dark ball as the dimmer star. Think about what your binary would look like from Earth as you conduct the experiment. When would you expect the brightness of the system to be greatest? How do you think the stars will be arranged when the system is faintest?

YOU WILL NEED
- *small bright ball*
- *large dark ball*

Viewing a binary from Earth
Put the balls on a table, and ask a friend to move the balls around a point between them, so that the small bright ball passes in front of and behind the large ball as you look from the side. This shows how a large dim star can obscure a smaller, brighter one.

Star birth

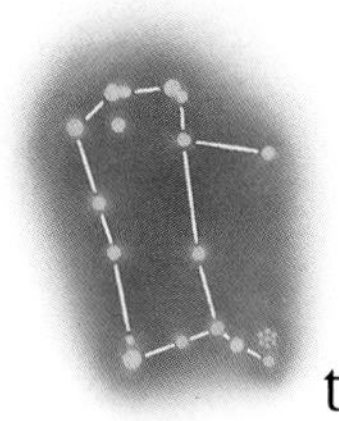

THE LIFE SPAN OF A STAR is very great indeed. Even the shortest-lived stars have a natural span of millions of years. Although we seldom see the stars actually in the process of changing as they age, we can see stars in different stages of development, so we can piece together a star's life story. A star is born when a huge, dense cloud of dust and gas that have collected over billions of years begins to collapse under its own gravity. Its central temperature rises to millions of degrees, and at these high temperatures the gas starts to undergo nuclear fusion reactions (p.154), giving out so much energy that the collapse stops—and the young star starts shining steadily.

EXPERIMENT

How stars are born

What causes a collapsing cloud of gas to get so hot? The answer is compression. Even air that is compressed inside a bicycle pump, as in this experiment, begins to get hot—though you could never pump hard enough to start nuclear fusion reactions!

YOU WILL NEED
- *bicycle pump*

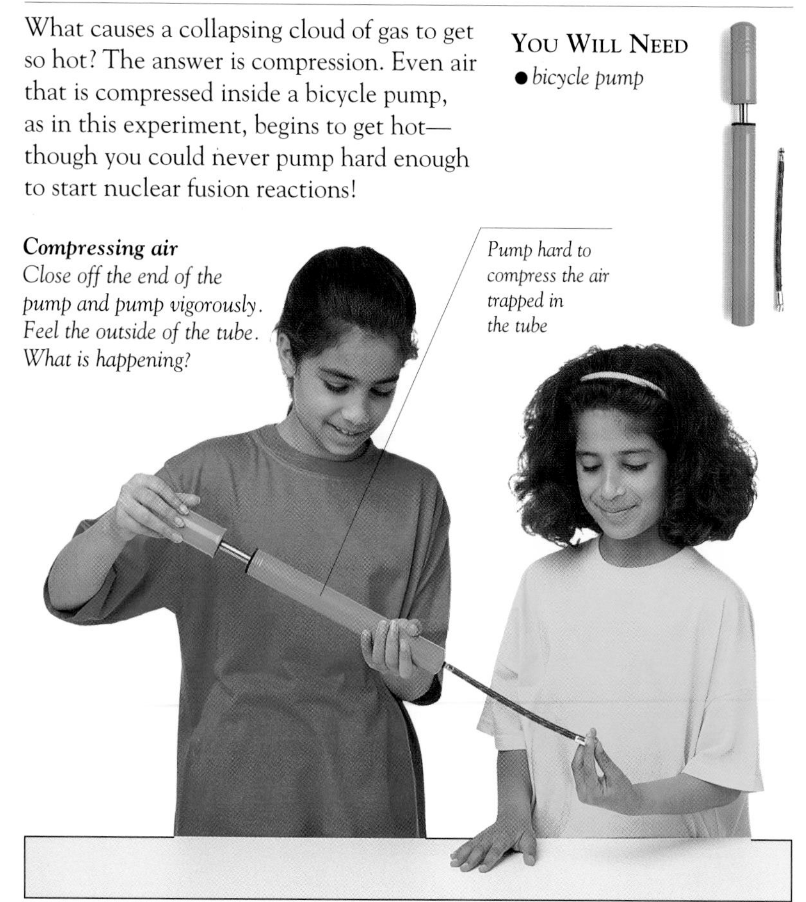

Compressing air
Close off the end of the pump and pump vigorously. Feel the outside of the tube. What is happening?

Pump hard to compress the air trapped in the tube

Finding the construction sites

You can identify some of the "construction sites" where stars (and possibly planets) have just come into being. These "nebulae" (the Latin word for "clouds") are some of the most beautiful objects in the sky. Some can be seen with the unaided eye, but they are best observed through binoculars or a small telescope. These diagrams show where to find bright nebulae (locate the constellations using the star maps, pp.148–151). Why not try sketching them?

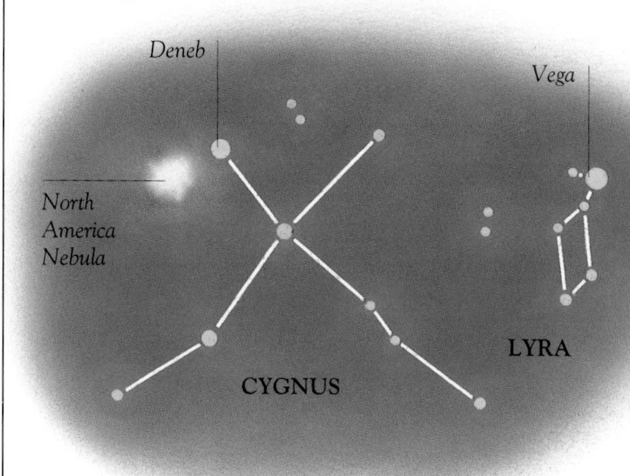

North America Nebula
Best viewed using binoculars or through a telescope, this nebula, in the constellation Cygnus, really does look like a map of North America. It even has its own "Gulf of Mexico," formed by a dark dust cloud. A large telescope may show the Pelican Nebula (too faint to be shown here). Both nebulae lie 1,500 light-years from us.

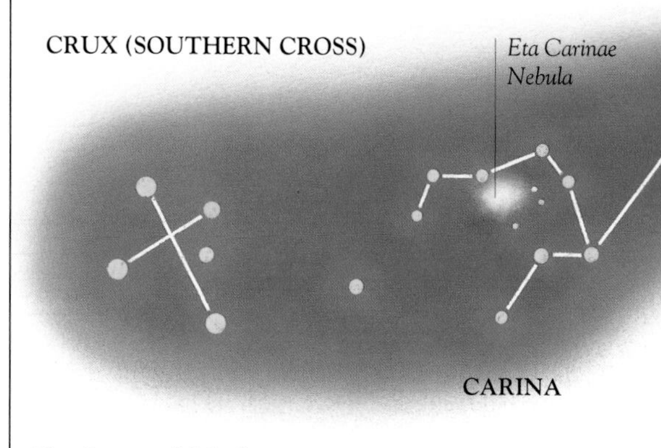

Eta Carinae Nebula
Lying in the Milky Way in the constellation of Carina, the brilliant Eta Carinae Nebula is visible to the unaided eye. With the help of binoculars, you can see long threadlike filaments within it. The nebula lies 9,000 light-years away and is one of the most brilliant in our Galaxy. The star in the center, Eta Carinae itself, is very unstable.

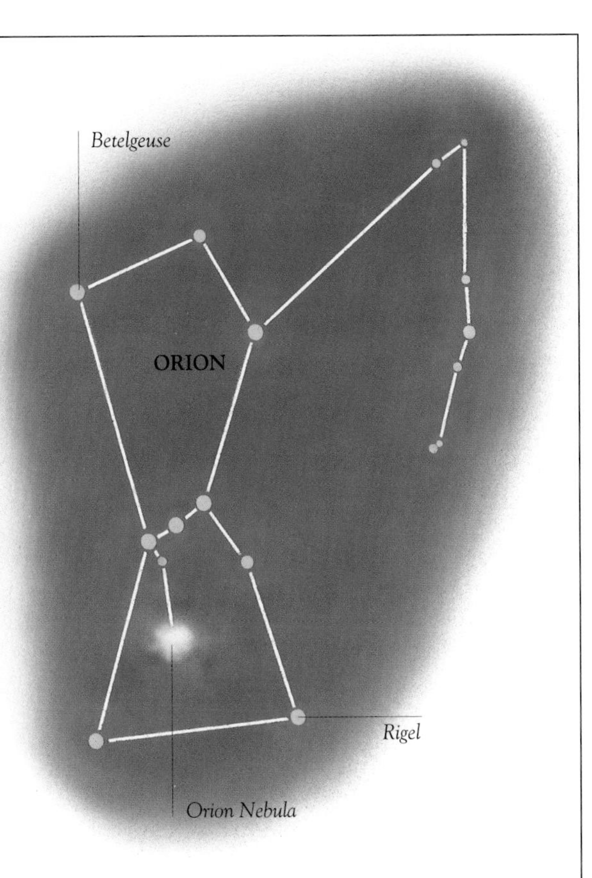

Orion Nebula
The best known of all the nebulae, the Orion Nebula is easily visible to the naked eye, stunning through binoculars, and amazingly detailed through a small telescope —which will reveal the "Trapezium," a cluster of four stars at the nebula's heart.

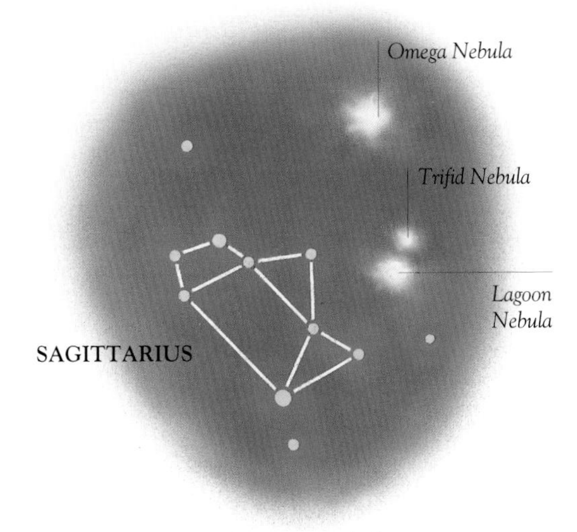

Trifid, Lagoon, and Omega Nebulae
All in the constellation of Sagittarius, in a very dense region of the Milky Way, these three nebulae are best seen from the Southern Hemisphere, preferably with binoculars or a telescope.

Heat-sensitive cameras

Special cameras mounted on satellites such as IRAS (below) help astronomers locate "hot spots" in the Universe—the places where stars are being born. The cameras are sensitive to infrared radiation (heat) and produce images called thermographs—color-coded "heat maps" like this image of two human bodies. Hot areas are white and red, shading to purple and black.

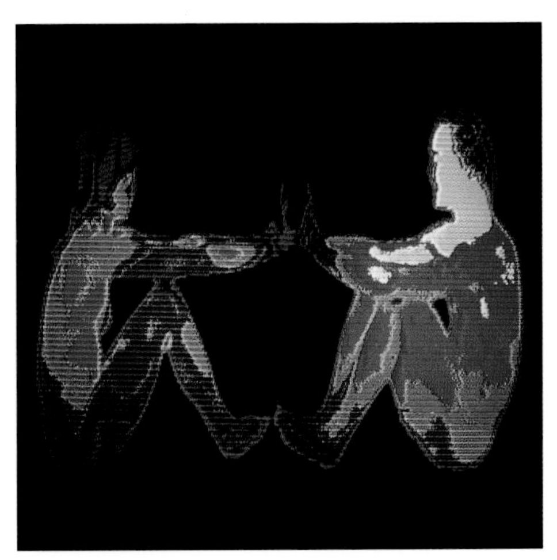

Searching for planets—the IRAS telescope

The IRAS satellite was launched in 1983 to detect cosmic dust using an infrared camera (IRAS stands for "Infrared Astronomical Satellite"). The Earth-orbiting telescope detected warm disks of dust around some stars—including Vega and Beta Pictoris. These may be planetary systems in the process of formation, made up of dust grains left over from the stars' birth.

Clouds of many colors

Through a small telescope, nebulae look simply gray and white. But if our eyes were sensitive enough, we would see many different colors. This photograph of the Orion Nebula has been enhanced to show the colors. The crimson color comes from hot hydrogen gas; the blues, from light scattered off tiny dust particles. Oxygen atoms can produce shades of green. The darker regions are thick bands of dust.

Star death

A STAR DIES when it runs out of fuel. Once a star has used up the stores of gas in its core, the nuclear fusion reactions that make the star shine can no longer take place. But stars die in different ways, depending on how massive they are. Most stars—including the Sun—use up their nuclear fuel slowly and live for billions of years. When they die, they do so quietly. Massive stars—those more than 10 times heavier than the Sun—live fast and furiously. After only a few million years, they develop unstable cores and literally end with a bang. Even when a star "goes supernova," as these explosions are called, it is not all destruction. In the heat and fury of the explosion, new elements are created. Thousands of years later, debris from the explosion triggers the birth of new stars.

The death of a sun

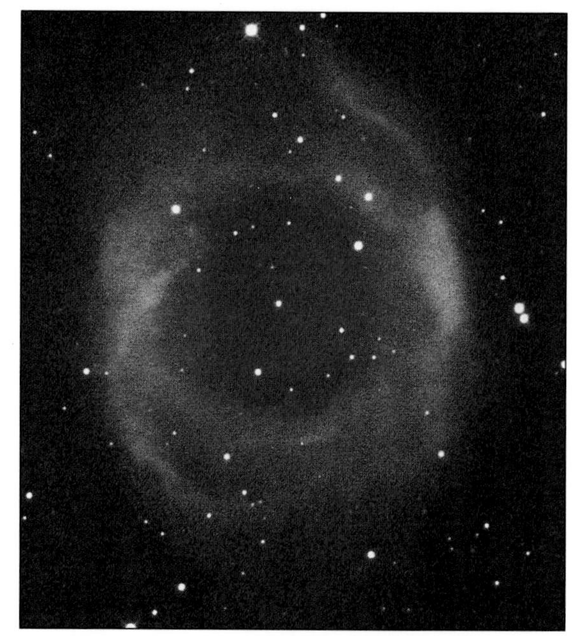

A small, average star like the Sun dies with a whimper. When it runs out of nuclear fuel, its central core shrinks and gets hotter, making the star's outer layers billow out. The star becomes a red giant. Its huge, inflated atmosphere is unstable, and the star eventually puffs it away into space. The puffed-away gas is called a planetary nebula because it looks similar to a planet when viewed through a telescope. Because of its shape, this planetary nebula (above) is called the Helix Nebula. At its center is the star's collapsed core—a white dwarf star that slowly leaks its heat into space.

EXPERIMENT

Stardust

If stars are hot, luminous balls of gas, how is it that they can produce solid matter—the essential raw material for planets and even life itself? When a candle burns, it produces hot gases that can be collected to form a solid. Ask an adult to help you perform this experiment, which will show you how something can be created out of thin air.

Something from nothing

Ask an adult to light a candle and then lower a saucer into the flame for a few seconds. Watch what happens. The saucer will become coated with a black deposit of soot, or carbon. The candle flame is very hot, and its yellow color shows that it is burning carbon. The carbon is not visible because its particles are extremely small—some have only a few dozen atoms each, just slightly bigger than the particles of a gas. When the flame meets the cool saucer, the carbon condenses, forming soot. In the case of red giant stars, hot carbon gas condenses in the cold of space to form clouds of soot, which later float away.

Lower the saucer into the flame

Adult supervision is advised for this experiment.

YOU WILL NEED
- *candle*
- *matches*
- *saucer*
- *oven mitt*

EXPERIMENT

Why stars explode

The most massive and luminous stars die by blowing themselves up. A star explosion is called a supernova, and it is an incredibly powerful affair: an exploding star can shine as brightly as an entire galaxy of thousands of millions of stars. A supernova actually starts with a collapse. Under extreme conditions of heat and pressure, the star's core—its "nuclear reactor"—breaks down. With no energy flowing out, it shrinks dramatically. Eventually it becomes a solid ball made up of neutrons—the densest form of matter—that is only a few miles wide. Why does the rest of the star explode? This experiment demonstrates what happens.

You Will Need

- *large ball*
- *small ball*

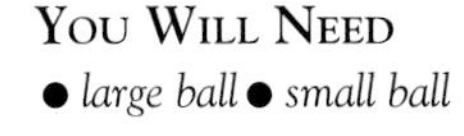

On the rebound

In a large open space, hold the small ball on top of the large ball and drop the two from a height of about a yard. Are you surprised by how high the small ball bounces? This is because it has bounced off the solid surface of the larger ball. In a similar way, the gas that follows the collapsing core of a star about to "go supernova" eventually rebounds off the core's solid surface. The bounce is enough to lift off the upper layers of the star, and they explode into space.

How to make gold

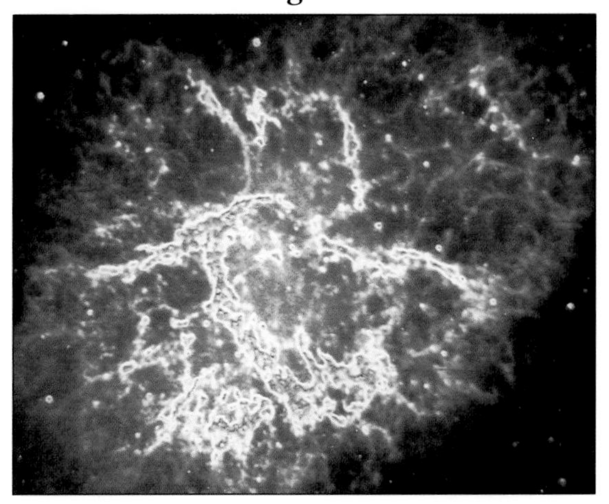

Where do heavier elements such as gold, silver, and uranium come from? They have come from supernovae, such as the one that produced the Crab Nebula (above). The extremely high energy conditions in a supernova create an environment like that of a particle accelerator, where atoms smash together. New atoms form, but in small quantities, which is why elements heavier than iron are so rare.

Observing supernovae

Supernovae are very rare—perhaps only three occur in our Galaxy every century (and most of these are hidden by dust in space). But we can see supernovae exploding in other galaxies. Some supernovae are discovered by amateur astronomers who patiently scrutinize dozens of galaxies every night. Most are so distant that they are visible only with a telescope, but one exception appeared in February 1987. This supernova (lower right, above) exploded in the nearest galaxy to ours, the Large Magellanic Cloud. It was 250 million times brighter than the Sun, and it was visible to the naked eye for several months.

Pulsars and black holes

PULSARS AND BLACK HOLES were unknown to us until a few years ago. Astronomers then found that strange things can happen to the core of a star after a supernova explosion. If the core is between $1\frac{1}{2}$ and 3 times heavier than the Sun, the constituents of its atoms collapse, and the core becomes a ball of neutrons—a neutron star, or pulsar. If the core is more than 3 times the Sun's mass, it just keeps on collapsing, growing denser and denser. Eventually, the gravity at its surface becomes so strong that it prevents light from getting away. Because light cannot escape, it becomes totally black. Nothing can travel faster than light, so anything falling in is trapped forever in this ultimate hole. No one knows what lies at the center of a black hole, but astronomers believe it could be a "singularity"—a point of infinite density.

What if you fell into a black hole

Imagine falling into a black hole feetfirst. The first thing you would notice would be a slight pull on your feet, which, being nearer the hole, would feel its gravity first. As you approached the hole, gravity would pull on your feet ever more strongly, and your body would start to stretch. You would be pulled into a long, thin tube—a process that is for obvious reasons called "spaghettification"—before you finally disintegrated.

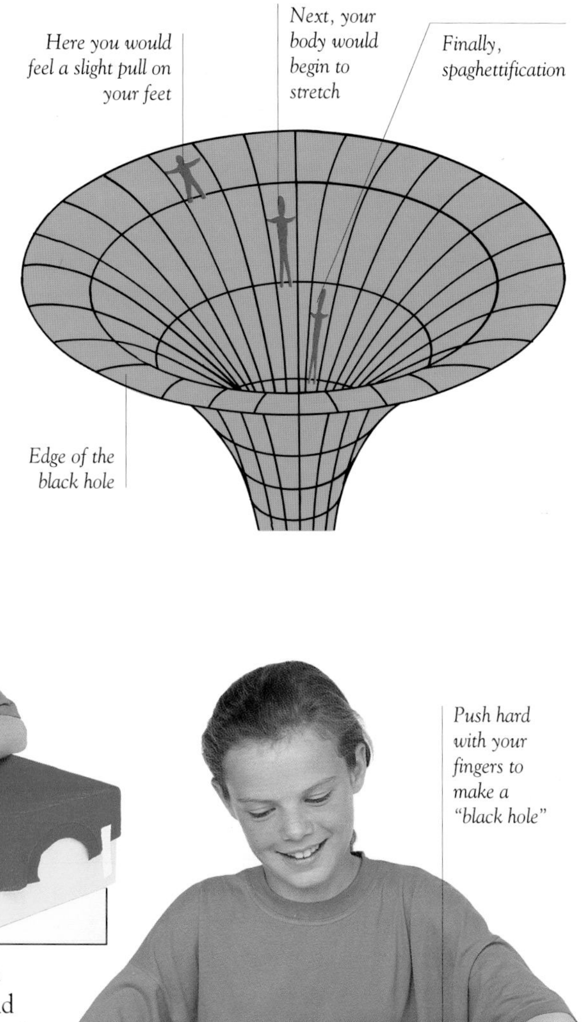

EXPERIMENT

Making a black hole

One way of looking at black holes is to imagine space and time as a flexible sheet and gravity as the shape of the sheet. Putting a lightweight object on the sheet makes little difference to its shape. But if you put a massive object there, the sheet bends. If you now add a lighter object to the sheet, it will fall into the "pit" of the massive object. This is another way of saying that massive objects have stronger gravity than less massive ones. You can make your own "black hole" in a sheet of "space-time" using an old T-shirt and a box.

YOU WILL NEED

- *old T-shirt* ● *scissors*
- *large cardboard box*
- *marble* ● *tape*

1 CUT AN OLD T-SHIRT into a single flat piece of cloth, and fix it tightly over the open end of the box with tape.

2 STICK YOUR FINGER in the middle of the T-shirt, pressing hard or lightly to represent objects of different masses. Put the marble near the edge of the box, and let it go. How does the marble move (if at all)? How does it compare to what you would expect if your finger were exerting a gravitational pull? Imagine what a super-dense, massive object you would need to create a tear in space-time.

EXPERIMENT

Making a pulsar

When a star explodes as a supernova, its core—a tight ball of neutrons—may survive the blast. Newly formed "neutron stars" spin very rapidly, often several times a second. Some, called pulsars, appear to flash beams or pulses of radiation that can be picked up by radio telescopes. We detect a flash each time the neutron star's rotation reveals a bright "hot spot" on its surface. So you have to be in exactly the right place to detect the flashes. This means our Galaxy must contain many more pulsars than the ones already known: we just don't see the others flashing. Below, you can make your own flashing "pulsar."

You Will Need

- *table-tennis ball*
- *2 batteries (1.5 V)*
- *thick tape*
- *flashlight bulb and holder*
- *wire*
- *modeling clay*
- *wooden skewers*
- *scissors*
- *screwdriver*

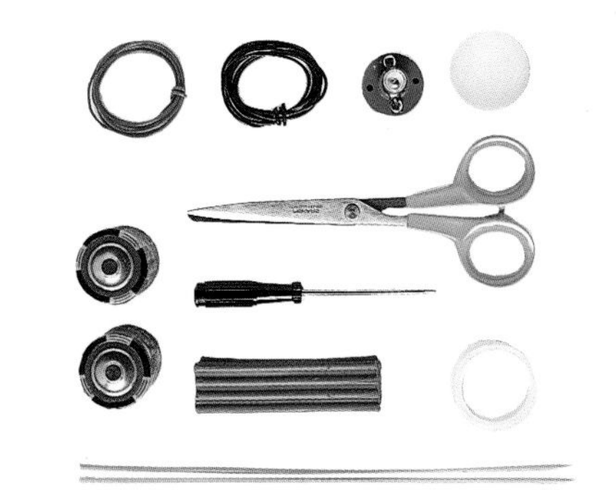

The discovery of pulsars

Jocelyn Bell and Tony Hewish (below) stand inside the radio telescope in Cambridge, England, with which they discovered the first pulsar in 1967. Their "telescope" consisted of 2,048 wire antennas—each line an FM radio antenna – strung out over a field almost 4 acres (2 hectares) in area. The antennas collected weak radio signals from space, including the first regularly repeating pulses from neutron stars. The first pulsar flashed once every 1.337 seconds.

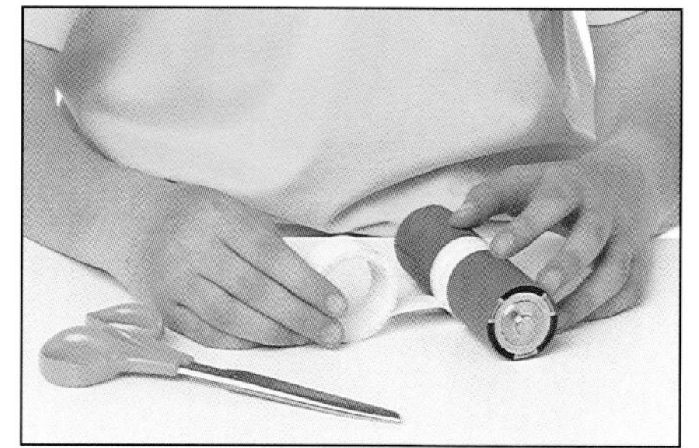

1 Take two 1.5-V batteries, and tape them firmly together. Be sure they fit end on end, so that the positive end of one battery fits into the negative end of the other battery.

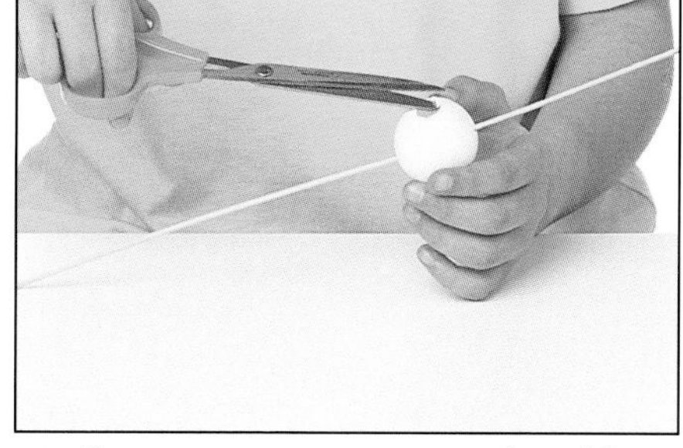

2 Stick two skewers into the table-tennis ball so that they are opposite each other. Then, between the skewers, cut a hole in the ball that is big enough to fit over the flashlight bulb.

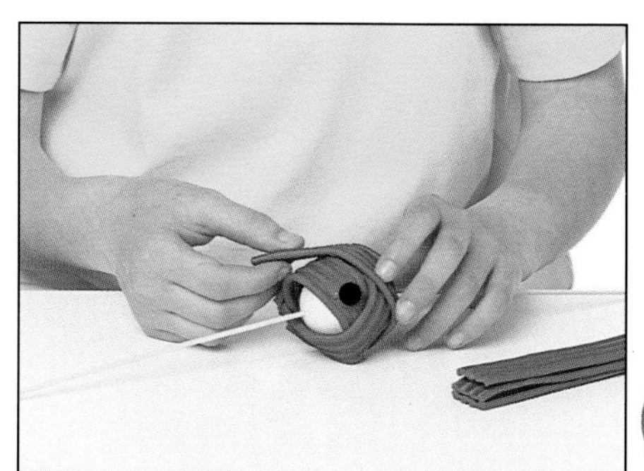

3 Cover the ball completely with modeling clay. Smooth the clay so that there are no gaps. Then, remove the two skewers, so that there are two tiny holes through the clay and ball.

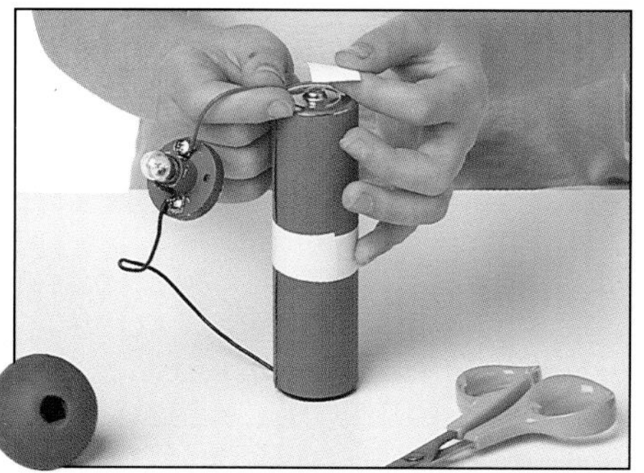

4 Finally, connect the bulb to the batteries, using wire and tape, and stick the ball firmly on the bulb. No light should escape, except from the two holes. Now darken the room.

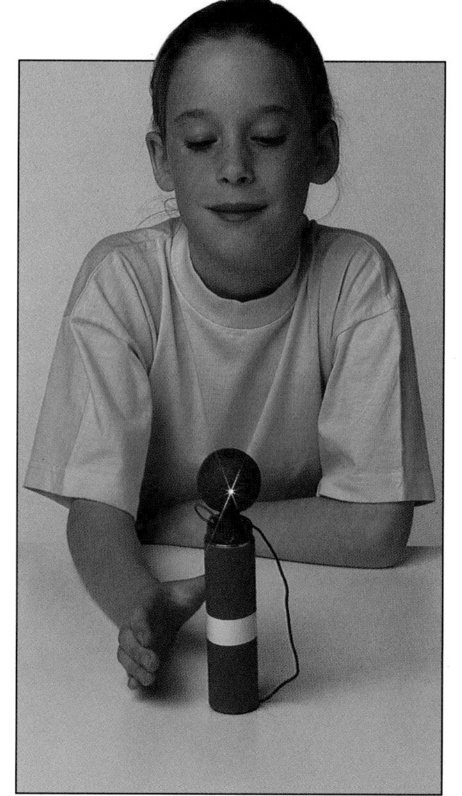

5 Slowly spin the "pulsar." As the holes pass your line of sight, you will see a flash of light. By watching how quickly the flashes repeat, astronomers work out the rotation rate of pulsars.

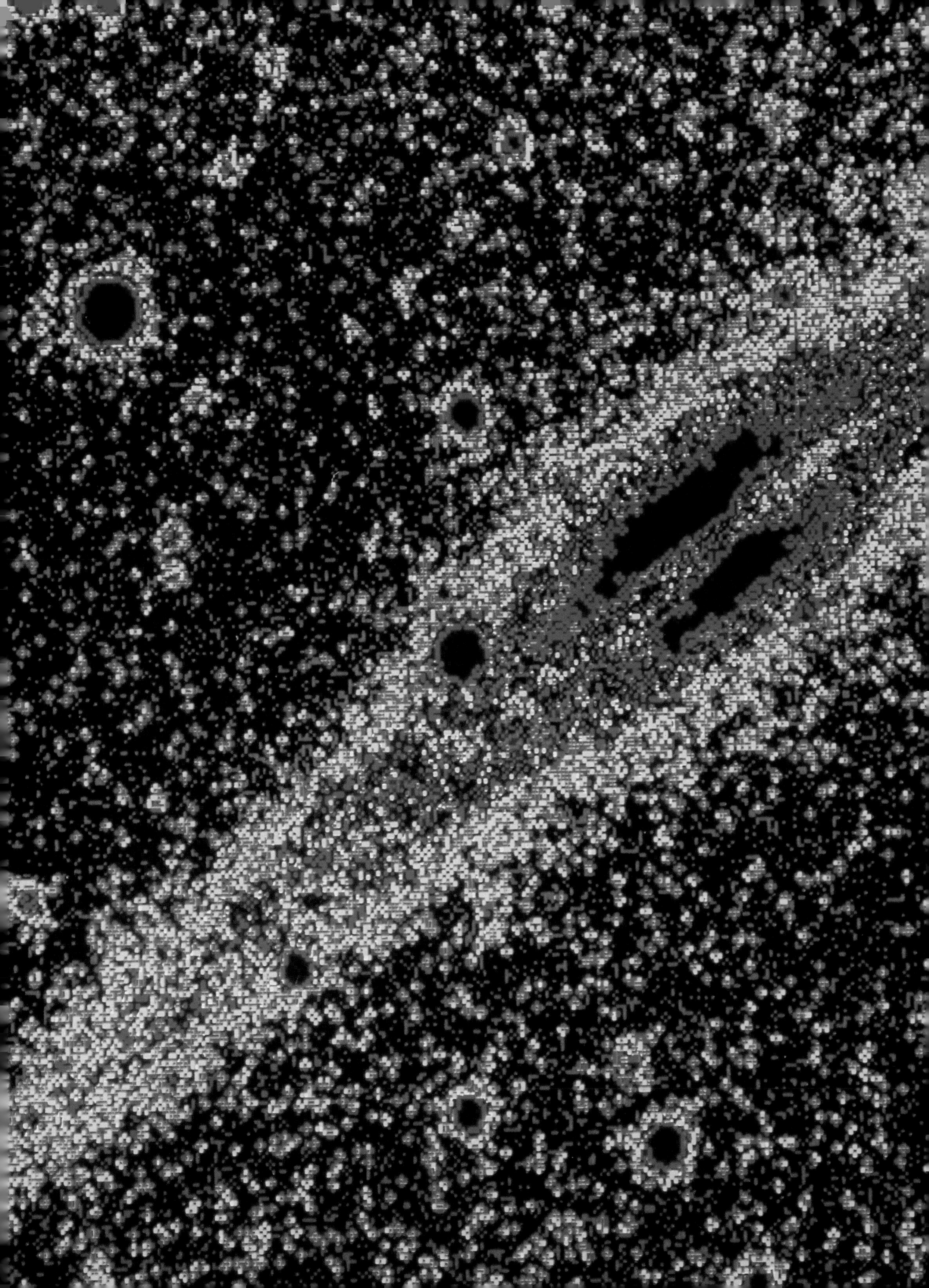

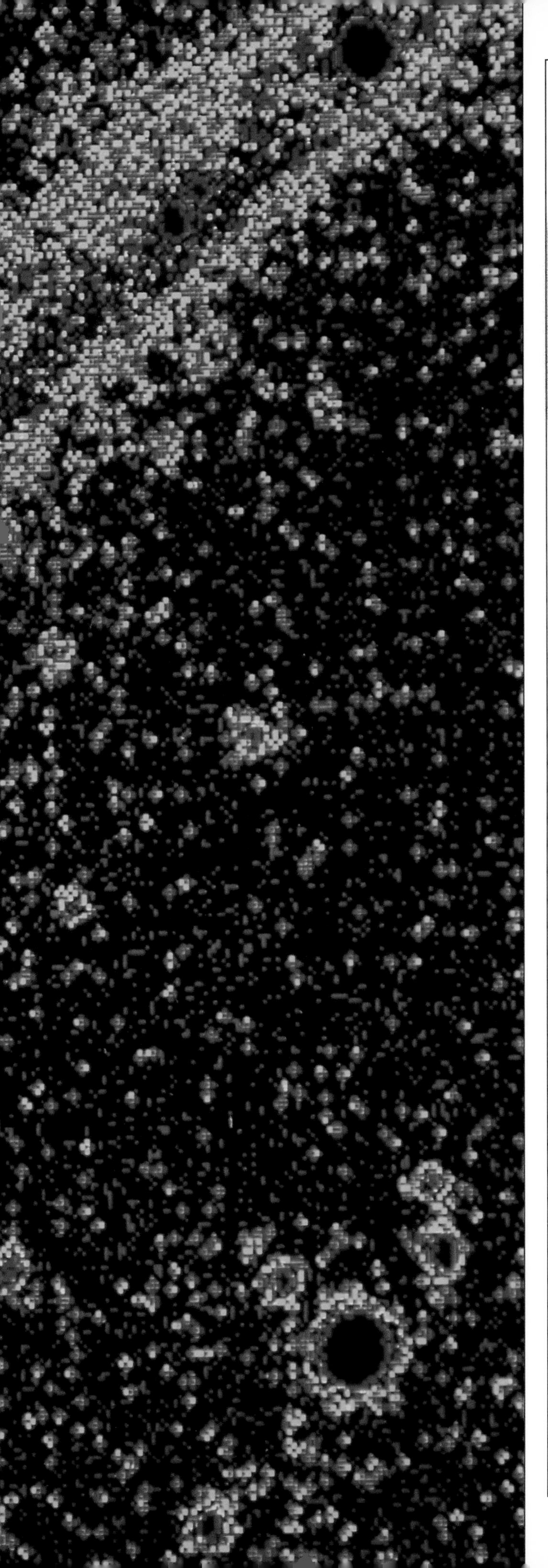

The COSMOS

Star cities

Our Universe is made up of many different "star cities," or galaxies. There are spiral galaxies (left) like our own—the computer-generated colors in this image indicate the brightness of the stars. Cygnus-A (above) is another type of galaxy, in which massive explosions have produced two huge clouds of gas.

GALAXIES ARE THE BUILDING blocks of the Cosmos. Gathering together by the hundreds and thousands into clusters and superclusters, galaxies form giant threads around empty space. But where did they come from and how were they formed? How did the Universe begin? And is there life somewhere out there?

GALAXIES AND BEYOND

THE STARS WE CAN SEE are just a small proportion of the 200 billion that make up our Galaxy, the Milky Way. A view from outside would be stunning: a huge slowly turning pinwheel 100,000 light-years wide. Curving spiral arms studded with brilliant blue stars and billowing nebulae spring from a smooth central hub of older, yellow stars. But if you were to search for the Sun, you would never spot it: like any average star, it merges into the combined glow of millions of others.

Living inside the Galaxy as we do, it is difficult to see the forest for the trees. However, the misty band we can see running through our skies gives some feel of the Galaxy's shape and extent. Sweep the band slowly with binoculars on a clear, dark night, and you will discover that it is made of countless stars. Because the Galaxy is flattened—it is shaped a bit like a fried egg—the farther you look, the more the stars appear to crowd into a band. This is why our Galaxy is called the Milky Way.

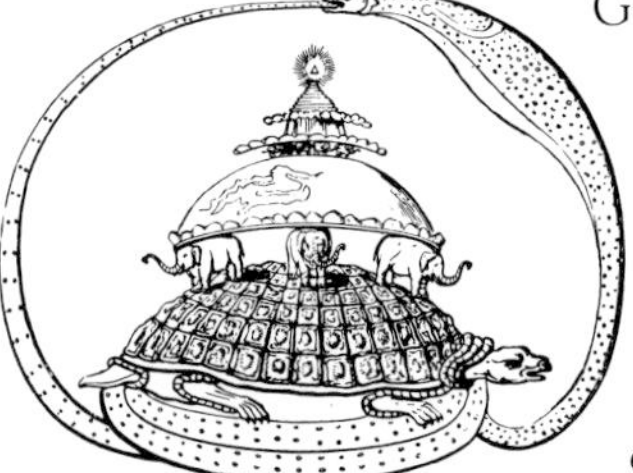

Hindu Universe
In early Hinduism, the egg-shaped Universe was symbolized as a snake (to represent eternity) encircling a mountain (paradise) on the Earth. Below the Earth was the underworld.

Inside our Galaxy

If you have a clear horizon, you will notice that the Milky Way itself looks brightest in the constellations of Scorpius and Sagittarius, where we are looking toward the galactic center. The Sun lies some 25,000 light-years —two-thirds of the way out— from the center, which consists of a rounded hub of older stars. Astronomers have recently discovered that there is more to the "nucleus" than that. Its very central regions are hot and bright, and there is a lot of fast-moving gas.

Our Galaxy
The Milky Way was photographed at the Las Campanas Observatory in Chile with a special camera that records the entire sky. Overhead is the center of the Galaxy, largely obscured by dark clouds of dust.

One reason why the events at the galactic center have only recently been noted is because our view is blocked by dust scattered between the stars. Astronomers cannot see the nucleus directly because the light from the Galaxy's core is dimmed over a million million times. Instead, they must study it with radio and infrared telescopes. These instruments pick up the longer-wavelength radiation that penetrates the interstellar dust. This obscuring matter is the raw material of future generations of stars. Although it is very thinly spread, the gas and dust have all the time in the Universe on their side. Over millions of years they collect together in enormous clouds, which eventually collapse to spawn stars. The gravitational pull of the spiral arms can squeeze the clouds and help star formation, so the newly born stars are strung out along the spiral arms.

Radio telescope
Antennas like these form part of the Very Large Array radio telescope in New Mexico. Linked electronically, these dishes provide a detailed view of distant objects that naturally emit radio waves, such as exploding galaxies and quasars.

Away from the bright lights of the Milky Way is the galactic halo— a vast, barren region surrounding the whole Galaxy. This was the first part of the Galaxy to be born, and it contains its oldest stars. There is more to the halo than you would think. The rate at which our Galaxy rotates suggests that there has to be a lot of matter in the halo, which is exerting a gravitational pull on the Galaxy's stars. As this matter is not immediately obvious, it must be dark. But what could it be?

47 Tucanae
Lying in the constellation Tucana, this bright globular cluster is just visible to the naked eye. It contains over a million stars, although only a few hundred are visible in this photograph through a telescope. Globular clusters are the oldest objects in our Galaxy.

Large Magellanic Cloud
The nearest galaxy to the Milky Way is the Large Magellanic Cloud, about 170,000 light-years away from the Earth. It contains many nebulae, including the huge Tarantula Nebula (top left)—so called because it resembles a large spider.

Galaxies like our Milky Way are probably made up mainly of this "dark matter" (pp.144–145), which may consist of enormous numbers of objects smaller than a star, but larger than a planet.

Beyond our Galaxy

Astronomers estimate that there are 100 billion other galaxies. Some of them are spiral in shape like our own; others have no well-defined shape and are called "irregulars." Both have the same mix of old and young stars, with large quantities of gas poised to form new stars. Elliptical galaxies, another type, are featureless round or oval balls consisting entirely of old stars. Some galaxies are hotbeds of violent activity. In the 1950's astronomers discovered "radio galaxies," which emit vast amounts of radio waves. In 1963 they found the first quasar—the exploding heart of a distant galaxy so brilliant that it outshone the galaxy itself. And in 1983 the Infrared Astronomical Satellite found some "starburst galaxies," where sudden outbursts of star formation produce enormous amounts of radiation in the form of heat.

Most galaxies live in clusters, held together by their gravity. There are even clusters of clusters —superclusters. Clusters tend to group around the edge of a supercluster. The effect is of filaments of galaxies enclosing voids of space. On a large scale, the texture of the Universe is rather like Swiss cheese.

M82: starburst galaxy
Astronomers once thought M82 was an exploding galaxy, like a quasar. It is now believed the disruption in its center has been caused by a sudden rash of star births. The radiation from this starburst has blown gas out into space.

Big Bang

Most astronomers today believe that the Universe began in a huge explosion called the Big Bang. Even today, all the galaxies are rushing apart from that explosion, and "background radiation" from the Big Bang fills all of space. What was there before the Big Bang? Nothing at all: both space and time were created in that instant, and the Universe has been expanding ever since. Will it expand forever —or will it collapse one day? The fate of our Universe depends on how much matter it contains, including the mysterious "dark matter." If there is enough matter, then the Universe has good "gravitational brakes"— the expansion will slow down, stop, and then reverse to a collapse. In the far-distant future, our planet Earth may face the "Big Crunch." But this seems unlikely because if the masses of all the stars and galaxies are added together, and the vast quantities of dark matter are taken into account, the total amount of matter falls short of the amount that would be needed to "close" the Universe. Thus, the fate of the Universe appears to be one of continuous expansion, with the Universe becoming colder and emptier as the stars, planets, and galaxies slowly die. An ever-expanding Universe gives us all the time it is possible to have, and we can use that time to explore our Cosmos in greater and greater depth. In our search we may even come across the answer to the most fundamental and exciting question of all: Is there life anywhere else in the Universe?

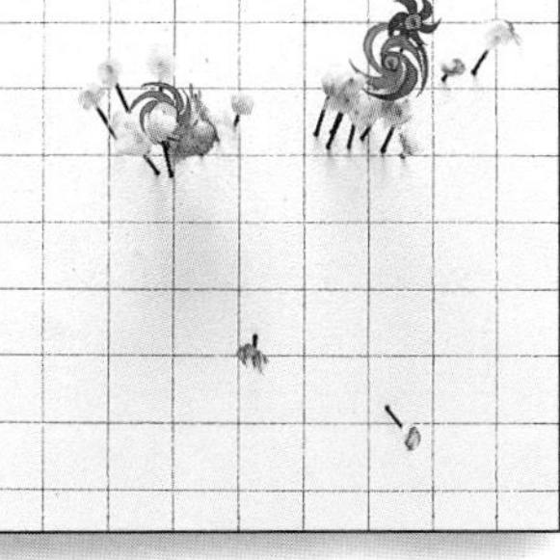

Our Local Group
The Milky Way is part of a group of galaxies. On page 139 you can make a model of our local cluster like the one above.

Arno Penzias and Robert Wilson
Two American scientists, Arno Penzias and Robert Wilson, won the 1978 Nobel Prize for physics for their discovery in 1965 of the cosmic background radiation, a "hiss" of radio noise that fills the Universe. The best explanation is that the radio waves have lingered from the Big Bang with which the Universe began.

The Milky Way

THE MILKY WAY is our Galaxy—a "city" of 200 billion stars, including our Sun. Seen from above, the Galaxy would look like a gigantic pinwheel, 100,000 light-years wide. From the side, it takes on the shape of an imaginary "flying saucer" or disk. Our Solar System lies two-thirds of the way toward the edge in the Galaxy's spiral arms (p.134), where star birth takes place, so we are surrounded by our city's bright lights—energetic young stars and glowing nebulae. The farther you look into the Galaxy, the more the distant stars concentrate into a band: the softly shining Milky Way of the night sky.

EXPERIMENT

The band of the Milky Way

Our Solar System is situated in the Galaxy's spiral arms, so we never see the Milky Way's spiral shape. To us it looks brightest in a band around the sky in the directions where we are looking at the disk sideways, seeing through the greatest "thickness," so the stars seem most concentrated. You can achieve the same effect by using an illuminated sheet of acrylic plastic as the Galaxy and viewing it from different angles.

Adult supervision is advised for this experiment.

YOU WILL NEED
- *clear acrylic plastic*
- *tape*
- *flashlight*
- *aluminum foil*
- *scissors*

1 ASK AN adult to take a flat piece of clear plastic (a plastic ruler will do, but it is better to use a piece of plastic that does not have a beveled edge) and tape it across the top of a flashlight, so that at least one edge sticks out at the side. The flat plastic represents the thin, flattened disk of the Milky Way.

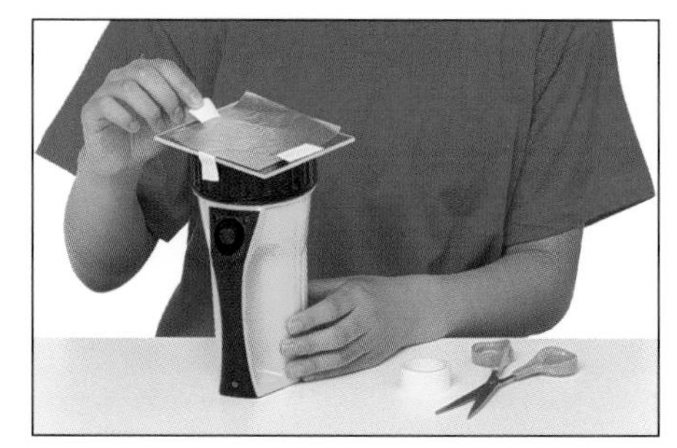

2 NOW TAPE A PIECE of aluminum foil over the plastic just big enough to cover up the light from the flashlight. Leave a small area of plastic around the edges uncovered.

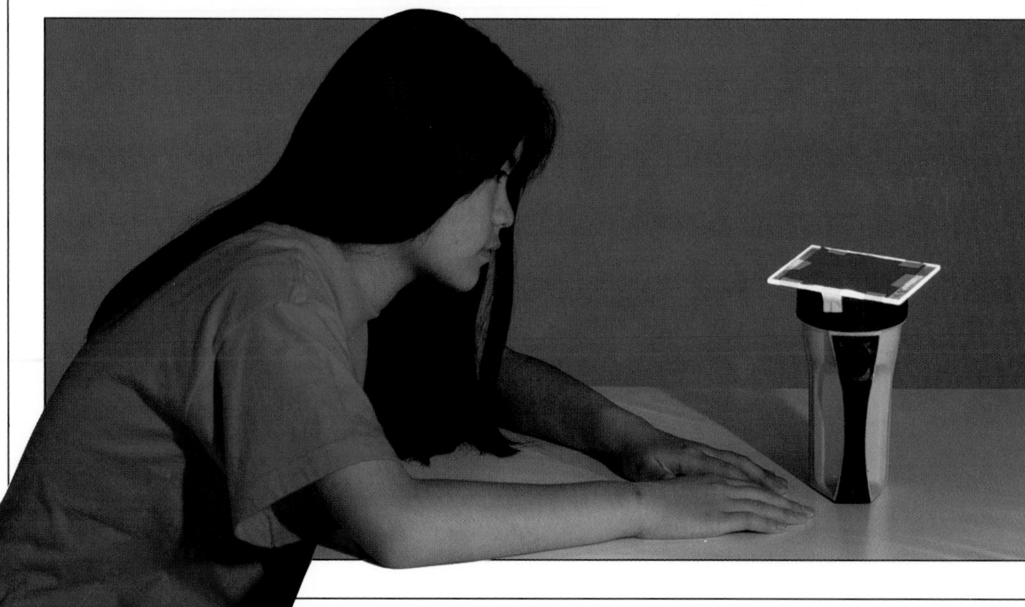

3 TURN OFF the lights in the room, and switch on the flashlight (so that your plastic "Milky Way" glows). Point the flashlight at your face, so that you are looking through the parts of the plastic that show around the foil. Now stand the flashlight on its end on a tabletop. Look at the edges of the plastic. Does it seem any brighter or fainter when viewed this way compared with seeing it from the top? How would the sky look if we were surrounded by glowing matter?

■ DISCOVERY ■

Jan Oort

Until quite recently, scientists believed that our Solar System lay at the heart of the Galaxy. In the 1780's William Herschel counted the number of stars in nearly 700 regions of the sky and, finding roughly the same density of stars everywhere in the Milky Way, concluded that the Earth was at the Galaxy's center. This view was confirmed in the 1920's by J. C. Kapteyn, who based his studies on the brightness and the motion of the stars. The Dutch astronomer Jan Oort (1900–92), one of this century's greatest scientists, proved otherwise. In the 1930's and 1940's he made many important discoveries about the size, shape, and nature of the Galaxy. He found that the Sun was not centrally placed and that the Galaxy rotates (and that the Sun takes about 220 million years to go around once). Using the newly invented radio telescope, Oort mapped the Galaxy's spiral structure. He was the first astronomer to guess at the existence of a cloud—now named the Oort Cloud—of millions of comets surrounding our Solar System.

Mapping the Milky Way

By observing our Galaxy during the year, you can work out roughly where the Sun is situated. Choose dark, clear nights, at a time when the Moon isn't visible and the Milky Way is high in the sky. Make a series of sketches over several nights in January and July. Mark the brightest stars, and pay particular attention to how bright and how wide the Milky Way looks. If we were in the center of the Galaxy, the Milky Way would look the same all the way around. Is this what you find? If the Solar System is at the outskirts of the Galaxy, then the Milky Way should look brightest and widest in the direction of the Galaxy's center. From your observations, in which constellation do you think the center of the Galaxy is located?

Use white chalk or a white crayon on black paper

What is the Milky Way made of?

Nobody could answer this question until 1610, when Galileo pointed the newly invented telescope to the skies. You can make the same discovery yourself. Go out on a very dark, clear night when the Moon isn't up and the Milky Way is high in the sky. Late evenings in July and January are ideal. With binoculars or a telescope, slowly sweep along the band of the Milky Way. What do you see? The hazy glow of the Milky Way is the combined light of thousands of stars. They seem tightly packed together, but while some of the stars are quite near us, others are many thousands of light-years away.

The structure of our Galaxy

IT IS DIFFICULT TO PICTURE the overall structure of our Galaxy, the Milky Way, because we live inside it and cannot get a view from the outside. Moreover, the space between the stars is filled with tiny grains of dust (each less than a thousandth of a millimeter wide), which absorb light. Dense clouds of dust can block out whole regions of the Galaxy. But by mapping our Galaxy using wavelengths longer than light that can get through the dust—such as infrared and radio waves— astronomers have found that the Milky Way has three distinct regions. A central "hub" of old stars is surrounded by a disk of young stars, gas, and dust, much of it bunched up in spiral arms. Surrounding the whole is a large empty region called the "halo," whose most obvious inhabitants are the oldest members of the Milky Way—the globular clusters.

EXPERIMENT

A model Milky Way

The Milky Way has three distinctly different parts. You can build a three-dimensional model to get an idea of what they are like and how they relate to each other. The center is a mass of old yellow-orange stars; stars are born in the dense dust clouds of the spiral arms; and globular clusters orbit far out in space.

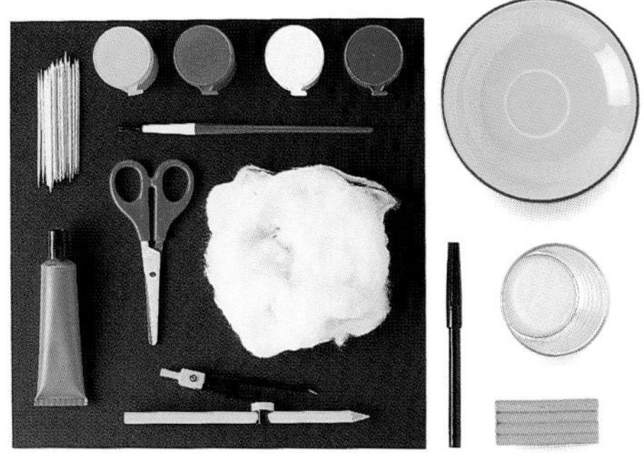

YOU WILL NEED

- *black poster board* • *compass* • *scissors*
- *poster paints* • *paintbrush* • *pen* • *cotton*
- *glue* • *orange modeling clay* • *toothpicks*
- *saucer* • *water*

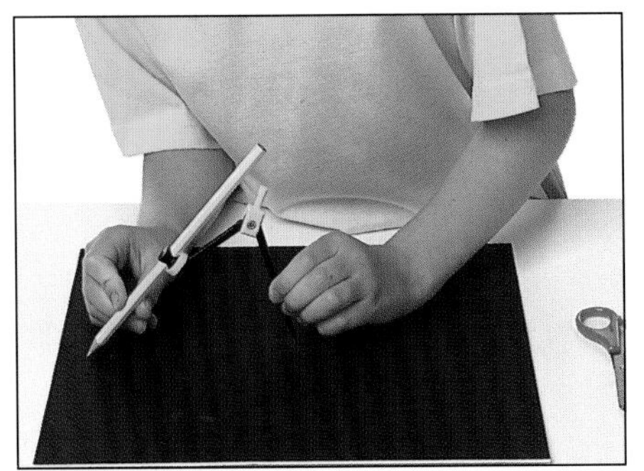

1 USE YOUR COMPASS to draw a circle 1 ft (30 cm) in diameter on a piece of black poster board. Then cut out the circle. This represents the disk of our Galaxy, 100,000 light-years wide.

2 PAINT A DOUBLE spiral shape as above. Dot the spiral arms with brown ("dust clouds"), pink ("nebulae"), and blue ("hot young stars"). Put an "X" (the "Sun") two-thirds of the way out on one of the arms.

3 MAKE TWO FLATTENED balls of cotton 3 in (8 cm) wide. Glue these in the center on both sides of the disk. Paint the balls yellow-orange. They represent the bulging hub of the Galaxy.

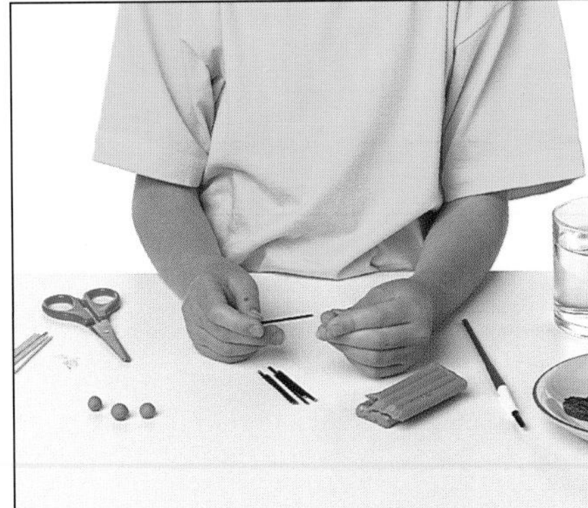

4 FOR THE FINAL COMPONENT, make balls of orange modeling clay 1/4 in (5 mm) wide. Attach them to toothpicks painted black. Each represents a globular cluster—a ball of very old stars.

5 SET THE "GLOBULAR clusters" around the "hub" and "disk" of your model Galaxy, with most toward the center and some farther out. This should give you an idea of the extent of the Galaxy's halo.

Spotting the Galaxy's fossils

The first parts of the Milky Way to be born were the globular clusters. Astronomers have found 140 globular clusters, each a ball of hundreds of thousands of old red stars. The stars at the center of a globular cluster are so dense that many would outshine the full Moon if we lived inside the cluster. Even thousands of light-years away, the brilliant globular clusters are easily visible with binoculars, your homemade telescope, or even the naked eye. These charts show you how to find some of the brightest globular clusters. The "M" numbers show that they were listed by the French astronomer Charles Messier.

M22
Located among the rich star clouds in the constellation of Sagittarius, M22 was the first globular cluster to be identified, by the German astronomer Abraham Ihle, in 1665. At 10,000 light-years away, it is one of the closest globular clusters to the Earth. It is just visible to the naked eye and is easily seen with the help of binoculars or a telescope.

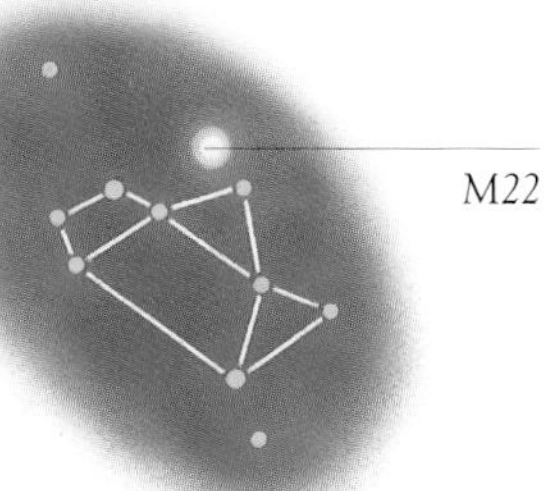

M13
Half a million stars make up this huge cluster in Hercules, 24,000 light-years away and visible to the unaided eye on a clear night. If the Sun were as far away as M13, it would be nearly 200,000 times too faint to be seen.

M13
HERCULES
SERPENS CAPUT
M5

M5
In the constellation of Serpens, M5 is one of the brightest clusters, visible—with binoculars—from the Northern Hemisphere.

47 Tucanae
A Southern Hemisphere cluster, 47 Tucanae lies 15,000 light-years away and at 10 billion years old is a relatively youthful cluster. One of the first to be identified, it can just be seen with the naked eye as a fuzzy "star" close to the Small Magellanic Cloud (p.125).

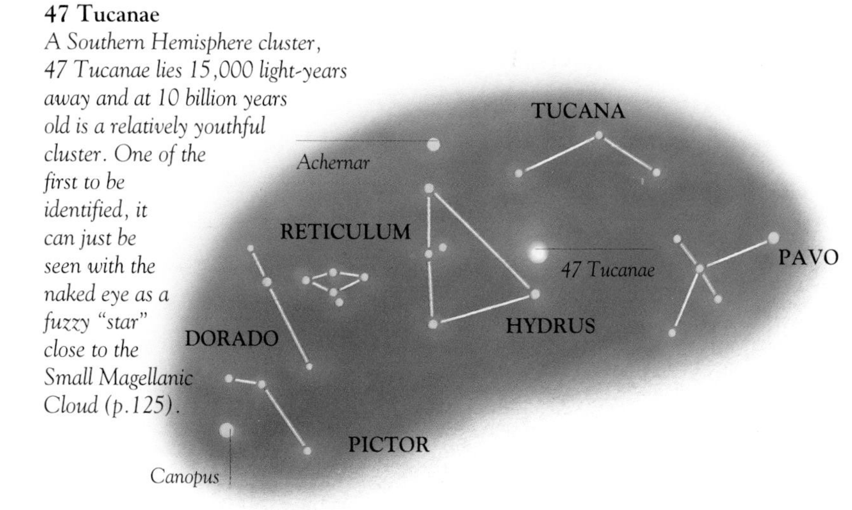

EXPERIMENT

An interstellar cloud

In the spiral arms in particular, tiny particles of dust block off the light from distant stars. Astronomers mapping the more distant parts of the Galaxy must use radio waves or infrared radiation, emitted by many objects in space. These are so long in wavelength that they are not blocked by the tiny dust grains. You can use a blanket to block off a radio's sound, just as interstellar dust clouds block off light. Note that certain sound waves still get through.

YOU WILL NEED
- *portable radio*
- *thick blanket*

Radio waves
Tune in to a rock music station, and wrap up the radio. The higher-frequency sounds (like cymbals) will be blocked, just like high-frequency light waves, but deep sounds with longer wavelengths will get through.

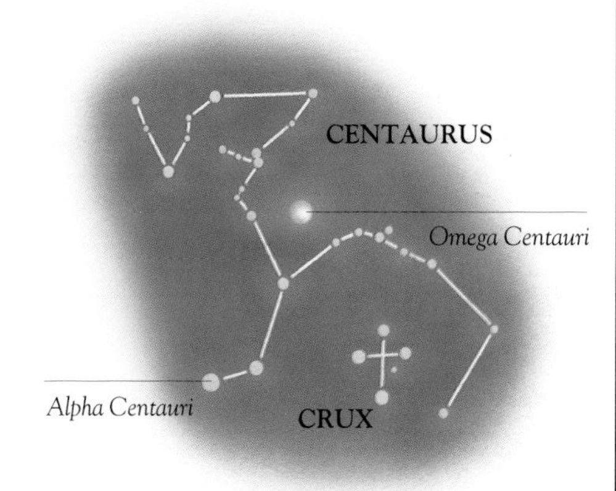

Omega Centauri
Omega Centauri is the largest and brightest cluster in the Galaxy and also one of the oldest. Easily visible in the Southern Hemisphere, it is two-thirds the size of the full Moon and distinctly bulges at its equator.

Galaxies galore

VAST THOUGH OUR Milky Way is, it is only one of an estimated 100,000 million galaxies that make up the Universe. Many of these are spirals like our own, with the same mix of old and young stars, plus the gas and dust that will form future generations. Similar to the spirals, but smaller, are the irregular galaxies, which have no particular shape. Their makeup is very similar to a spiral, and proportionally they contain even more gas to make future stars. The third main group of galaxies is the ellipticals. These contain only old red stars because, unlike other galaxies, they turned all their gas to stars at one time. They look just like balls of old stars, but they cover a huge range of sizes. The biggest and smallest galaxies in the Universe are ellipticals.

Galaxy spotting

On a clear night, a few galaxies are visible to the naked eye in addition to the Milky Way: the Large and Small Magellanic Clouds are prominent in the Southern Hemisphere, and the Andromeda Galaxy in the Northern Hemisphere. A pair of binoculars or a small telescope will reveal several more. Use these charts (below) to go galaxy spotting. You will need to go out on a very dark and clear night —with no moonlight and no trace of clouds. Sketch the galaxies carefully. Do they look like the photographs shown opposite? The human eye, even with the help of a telescope, is not nearly as sensitive as a long exposure on film, which is why the invention of photography has helped us learn so much more about galaxies. For example, where the eye sees only the bright center of a galaxy like Andromeda, a photograph will pick up fainter regions, such as its spiral arms.

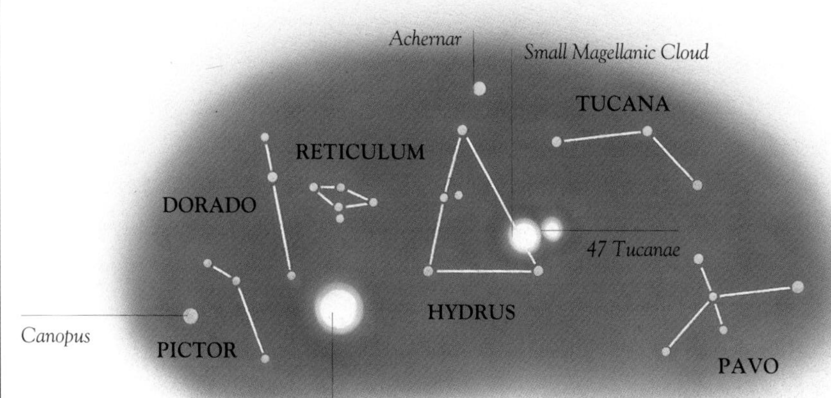

Large Magellanic Cloud (LMC)
This galaxy is a permanent sight in southern skies and, at 170,000 light-years from the Milky Way, is our closest galaxy. In 1987 a supernova exploded in the LMC and was visible for many months.

Small Magellanic Cloud (SMC)
This irregular galaxy and the LMC were first noted by Magellan's navigator on his round-the-world voyage in 1521—hence the names. The SMC has only 0.025 the mass of the Milky Way and is rich in gas for future star formation.

Andromeda
Visible as a misty oval blur on fall nights in the Northern Hemisphere, the Andromeda Galaxy lies 2¼ million light-years away. It is a spiral galaxy, even bigger than our own, with 400,000 million stars.

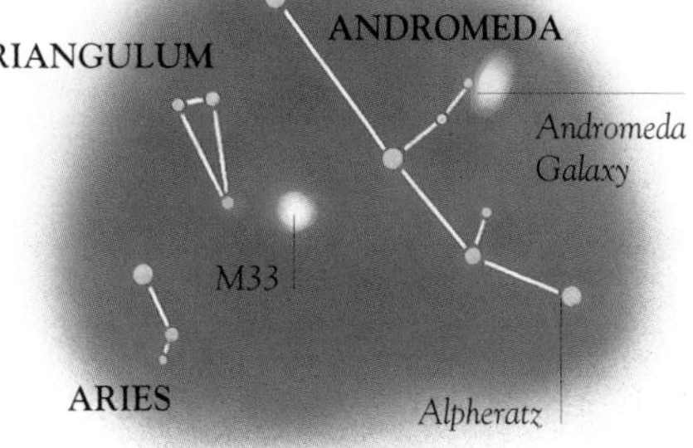

M33
This untidy spiral, 2.4 million light-years away, may be the most distant object visible to the naked eye, but it is much more easily seen with binoculars.

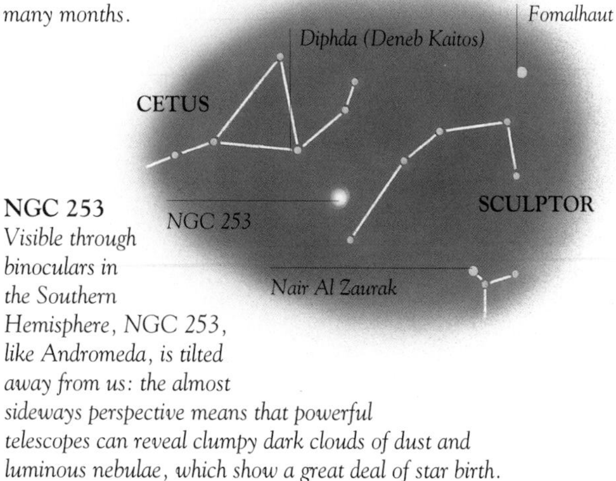

NGC 253
Visible through binoculars in the Southern Hemisphere, NGC 253, like Andromeda, is tilted away from us: the almost sideways perspective means that powerful telescopes can reveal clumpy dark clouds of dust and luminous nebulae, which show a great deal of star birth.

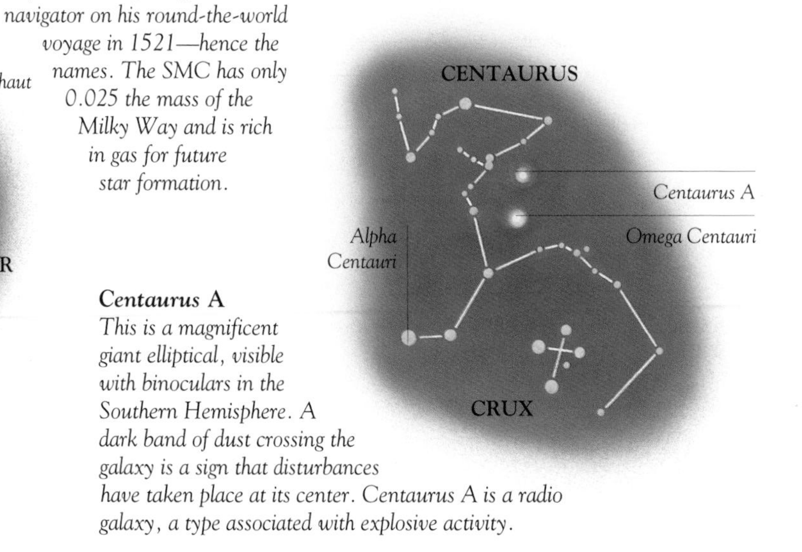

Centaurus A
This is a magnificent giant elliptical, visible with binoculars in the Southern Hemisphere. A dark band of dust crossing the galaxy is a sign that disturbances have taken place at its center. Centaurus A is a radio galaxy, a type associated with explosive activity.

EXPERIMENT

The puzzle of spiral arms

Astronomers believe that the spiral shape in galaxies such as the Milky Way can last for billions of years without winding up or fading away, even though individual stars swirl into it and then out again. From the results of the experiment below, do you think a galaxy's spiral shape can be maintained simply by the galaxy swirling around? If not, what kinds of force do you think could keep a galaxy in shape?

Adult supervision is advised for this experiment.

You Will Need

- *cup*
- *coffee*
- *spoon*
- *cream*

Swirling spiral
Ask an adult to make a cup of black coffee. Gently lower a spoonful of cream into it, and stir around and around in the same direction. The cream should form a spiral. Once the spiral is turning steadily, stop stirring. Does the spiral keep its shape? How long does it last?

Computer image

Astronomers very rarely work from conventional photographs. Photographs are attractive, but they do not provide precise measurements. Today, a scientist studying a galaxy captures its image electronically (with a device like a TV camera) and uses a computer to display it, as seen below. The computer has already processed, or "digitized," information about the galaxy, allowing a scientist to measure its properties accurately.

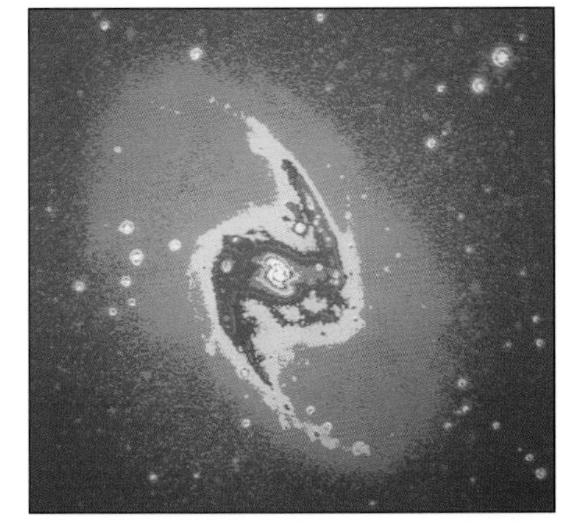

Color-coded data
In this computer image the different colors correspond to different intensities of light that the galaxy is emitting. In this case the computer gives us concrete and much more detailed information to back up what our eyes see when looking at an ordinary photograph.

Different kinds of galaxy

Just like people, galaxies come in many different shapes and sizes. You need a very powerful telescope, however, to make out the shape of a distant galaxy. In the 1920's American astronomer Edwin Hubble classified galaxies as spiral, elliptical, or irregular. Within the spiral group there are two distinct types: ordinary spirals and "barred" spirals. The latest evidence suggests that our own Galaxy, the Milky Way, is a barred spiral galaxy.

Spiral galaxy (M83)
Every spiral galaxy is different—astronomers who study galaxies all have their favorites. They differ in the size of their central hubs and the structure of their spiral arms—some are smooth and tightly wound, while others (like this one) are clumpy and loose.

Barred spiral galaxy (NGC 1365)
Instead of a round nucleus, some spiral galaxies have elongated central regions. The stars there form a long bar. The bar, made of millions of stars, rotates as a solid body—which is rather hard to explain.

Irregular galaxy (SMC)
This type of small shapeless galaxy has a lot of gas, from which countless generations of stars can form. Some very small galaxies (the dwarf irregulars) appear to have started serious star formation only recently.

Elliptical galaxy (M32)
This gas-free ball of old red stars can be round or oval, but it does not rotate as a whole. The stars follow their own orbits around the galaxy. Many ellipticals may have been stripped of gas by being too close to another galaxy.

Clusters of galaxies

GALAXIES ARE SOCIABLE. United by the bonds of gravity, they live together in groups or clusters that range in size from a handful of members to several thousand. Small clusters, like our Local Group, contain a mix of all kinds of galaxies. Giant clusters are dominated by gas-free elliptical galaxies—it seems that their gas was driven out at an early stage as they sped through the "atmosphere" between the galaxies. In some clusters, galaxies are so close that they interact with each other, drawing out curving streams of gas and stars. Galaxies may even swallow up one another. Astronomers have worked out that giant clusters must also contain large quantities of "dark matter"—something that has mass but no light.

Henrietta Leavitt

Henrietta Leavitt (1868–1921) was an astronomer at the Harvard College Observatory in Cambridge, Massachusetts from 1895. She discovered about 2,400 variable stars (p.115). When studying certain variable stars called Cepheid variables in the Small Magellanic Cloud (a galaxy that looks like a cloud to the naked eye), Leavitt realized that Cepheids of the same luminosity took the same time to go from maximum brightness back to maximum brightness again. By comparing the apparent brightness of Cepheids—dimmed by distance—in different galaxies, astronomers could then measure galaxy distances within the Local Group and beyond.

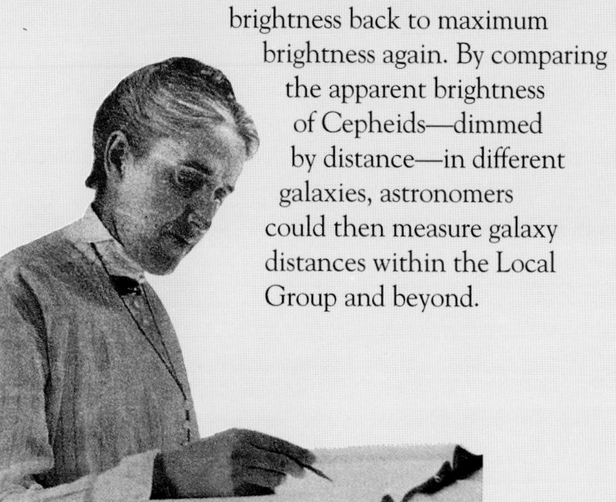

EXPERIMENT

The Local Group

The Milky Way and the Andromeda Galaxy are the two biggest members of a small cluster of galaxies in our neighborhood in space—the Local Group. The Local Group also includes the Magellanic Clouds, the spiral galaxy M33, and a few dozen tiny "dwarf galaxies." Using polystyrene as a base, you can make your own model of the Local Group, plotting the galaxies at the correct horizontal and vertical distances from one another.

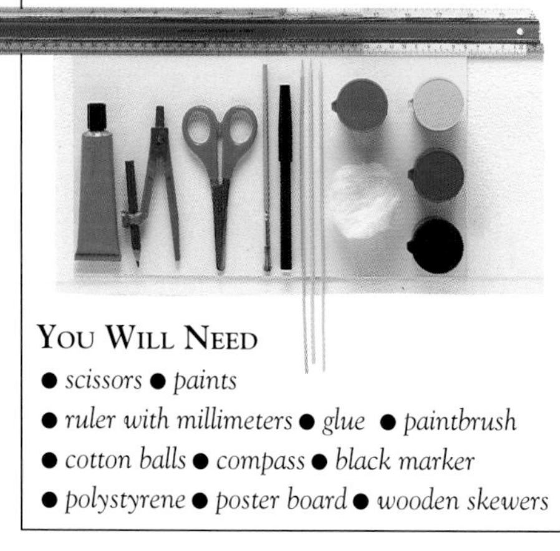

YOU WILL NEED

- *scissors* • *paints*
- *ruler with millimeters* • *glue* • *paintbrush*
- *cotton balls* • *compass* • *black marker*
- *polystyrene* • *poster board* • *wooden skewers*

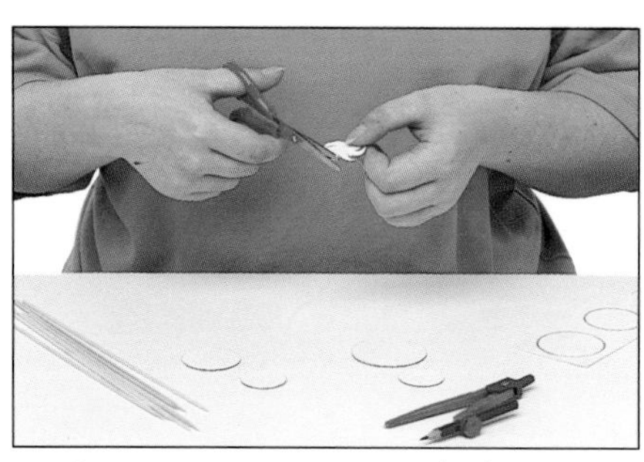

1 CUT POSTER-BOARD spirals 5 cm, 4 cm, and 3 cm wide to represent Andromeda, the Milky Way, and M33 respectively. Paint the spiral centers yellow and the arms blue.

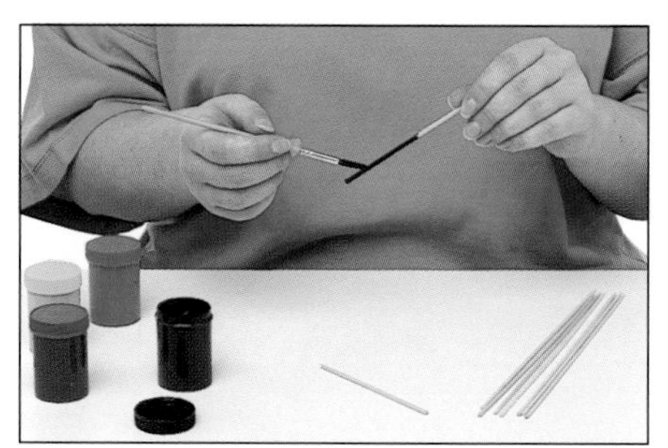

2 CUT 26 SKEWERS to length (see chart at right). Paint them black. Glue the spirals to three of them and cotton balls to the others. Dip the cotton-ball "galaxies" in blue or yellow paint.

3 TRIM A LARGE piece of polystyrene so that it measures 40 x 36 cm. Then use a black marker and a ruler to score the polystyrene with a grid of 9 squares by 10. Using the grid at right for reference, mark the position of each of the galaxies, and label them from 1 to 26.

EXPERIMENT

Cannibal galaxies

The galaxies living in giant clusters are spread anything but evenly. Usually, there are one or two "super-giant" galaxies, with a scattering of medium ones. This pattern has emerged because the galaxies move at high speeds and have cosmic collisions. When this happens, galaxies merge. These huge galaxies have such powerful gravity that they can "gobble up" many of the smaller galaxies. They are nicknamed "cannibal galaxies." This experiment uses magnetism to simulate gravity.

YOU WILL NEED

● *cork* ● *scissors* ● *pins*
● *magnet* ● *bowl* ● *water*

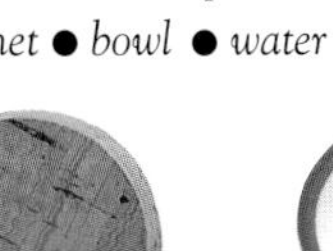

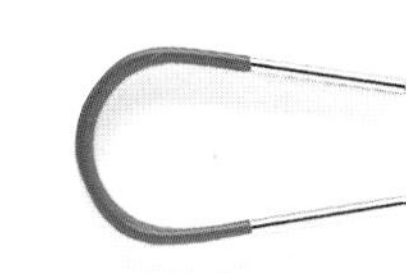

1 TAKE AN OLD CORK, and cut it into eight pieces that are more or less equal in size. Each of these pieces represents a single small galaxy.

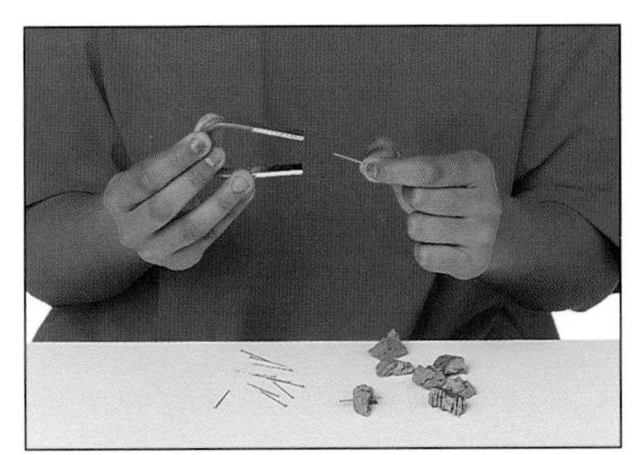

2 MAGNETIZE EIGHT PINS by stroking each several times with one end of a magnet, in the same direction each time. Put a pin through each piece of cork.

3 FINALLY, FILL a large mixing bowl about halfway with water. Drop in the corks, and spread them evenly throughout the water. Wait a few minutes. What happens? Is there any evidence of your cork "galaxies" merging or cannibalizing each other?

Skewer heights (in millimeters)

1. GR8 = 210	14. NGC 147 = 200
2. Leo I = 110	15. M32 = 190
3. Leo II = 105	16. NGC 185 = 200
4. Sextans = 95	17. NGC 205 = 170
5. Ursa Minor = 120	18. Andromeda = 180
6. Carina = 80	19. A-II = 165
7. Draco = 125	20. A-I = 175
8. Milky Way = 95	21. A-III = 170
9. S.M.C. = 80	22. M33 = 175
10. L.M.C. = 90	23. WLM = 70
11. Sculptor = 90	24. IC 1613 = 95
12. Fornax = 80	25. Pisces = 155
13. NGC 6822 = 80	26. DDO 210 = 70

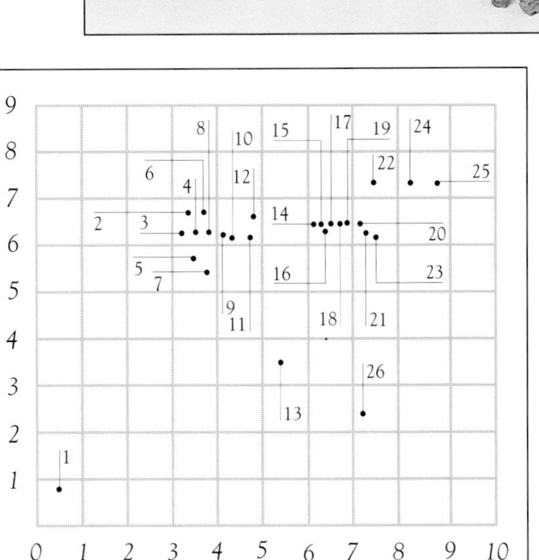

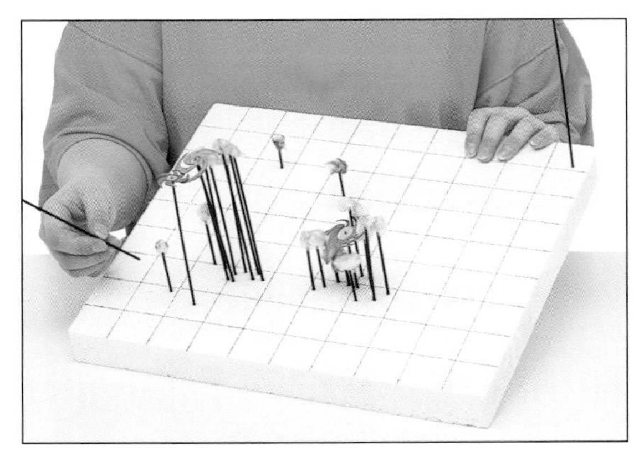

4 FINALLY, PLACE the skewers into the polystyrene in the correct position for each galaxy (see chart, above). This is how our Local Group looks from the outside. Do the galaxies spread evenly through the Local Group? If not, which galaxies do you think are "herding" up the smaller ones? What force do you think is responsible?

The Virgo cluster

Giant clusters of galaxies, like the Virgo cluster shown below, contain many large elliptical galaxies—some of which have "cannibalized" smaller galaxies living in the cluster. The Virgo cluster is the nearest giant cluster of galaxies to the Local Group. It lies 50 million light-years away and is 10 million light-years wide. With a small telescope—on a very dark night—you can see its brightest galaxies between Virgo and the star Denebola in Leo (see pp.148–151).

Quasars

BEYOND ALL THE NEARBY GALAXIES lurk the quasars—the most violent objects in the Universe. Through a telescope, a quasar looks like a faint star because it lies thousands of millions of light-years away. In reality a quasar packs the power of a million million stars into a region as small as the size of the Solar System. Quasars were discovered in 1963, and for many years they were a mystery. Now astronomers have found that each quasar lies in the heart of a very distant galaxy. Most likely, the quasar is a hot disk of gas swirling around a massive black hole in the galaxy's core. As well as shining brilliantly, a quasar can shoot out jets of electrons. These jets broadcast radio waves that astronomers on Earth can pick up with radio telescopes. In some cases, the quasar itself is hidden by surrounding clouds of dust, and all we can detect is the galaxy's radio emission. Astronomers call this kind of object a "radio galaxy."

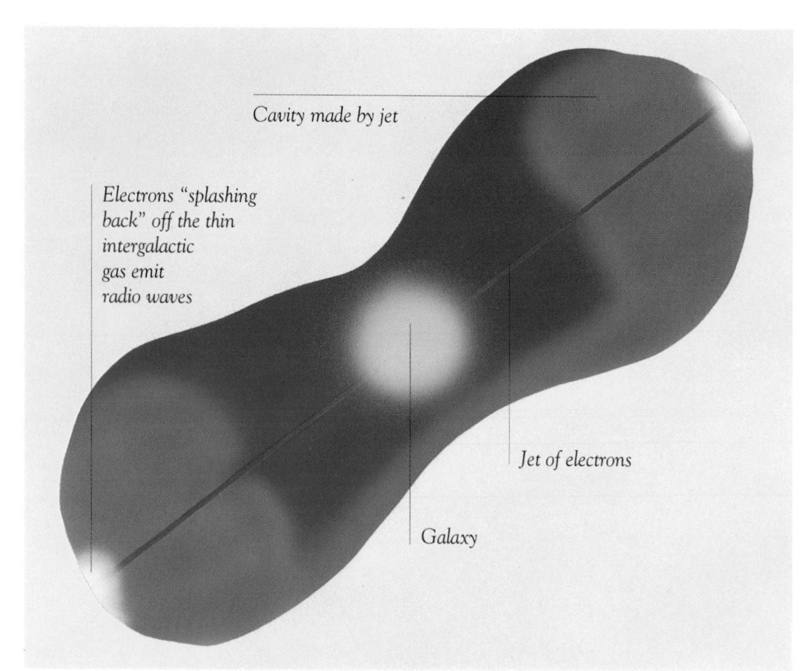

A radio galaxy

A quasar can produce two jets of electrons that push their way through the surrounding gas, excavating twin cavities on either side of the quasar. The cavities are filled with electrons, which produce radio waves. We can detect these cavities most easily in cases where we see the jets "frontally," and in this case the quasar itself is often hidden by dust clouds. Using a radio telescope, astronomers can then pick out the two glowing cavities, plus sometimes the jets (p.133). These radio galaxies can give out as much energy in radio waves as they do in light. The jets can be over 100,000 light-years long.

EXPERIMENT

The heart of a quasar

Astronomers believe that a quasar's heart consists of an enormous black hole surrounded by a whirlpool of infalling material. This whirlpool —the accretion disk—spins very fast, and it is the location of the quasar's violent activity. Here we investigate what happens to gas clouds and stars as they approach a black hole.

YOU WILL NEED

- *large mixing bowl*
- *cooking oil*
- *water*
- *spoon*

1 **FILL THE MIXING BOWL** about two-thirds full with water. Add a spoonful or two of oil so that several globules of various sizes form and float on the surface of the water.

2 **STIR THE WATER** until it dips in the middle, and watch the globules. Like matter streaming toward a black hole, the oil spirals in toward the center of the whirlpool.

EXPERIMENT

How small is a quasar?

When astronomers first found quasars, they discovered that these objects change in brightness in just a few months. These rapid changes told researchers that quasars must be extremely small—less than a light-year in size. The light from the front part of a quasar arrives before the light from the back of the quasar. The smaller the quasar, the more rapid any variations in brightness will appear. Here, you can use the same method to measure the size of your own "quasar."

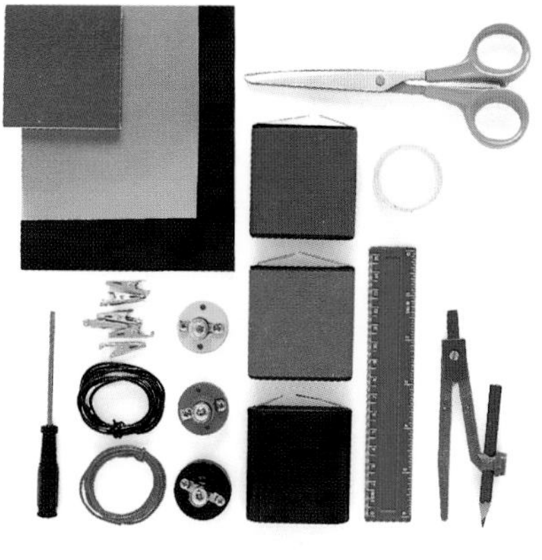

You Will Need

- *black and red poster board* ● *scissors* ● *ruler*
- *tracing paper* ● *3 flashlight bulbs with holders*
- *3 sets of connections and batteries*
- *screwdriver* ● *compass* ● *tape*

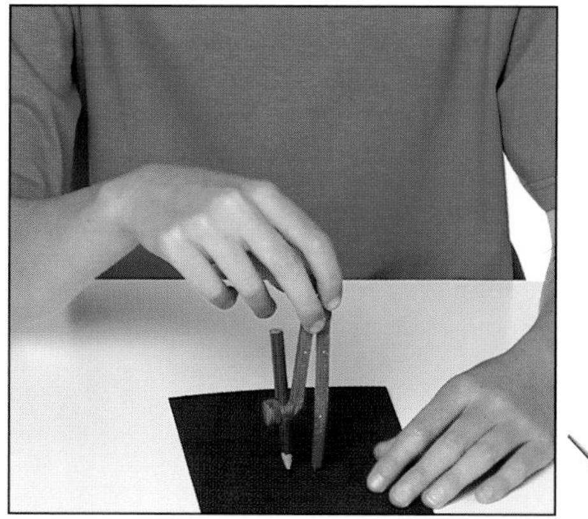

1 TAKE THREE PIECES of black poster board, and fold each in an L-shape. In the vertical surface of each cut a round hole of a different size— 1/4 in (5 mm), 1/2 in (1 cm), and 1 in (2 cm).

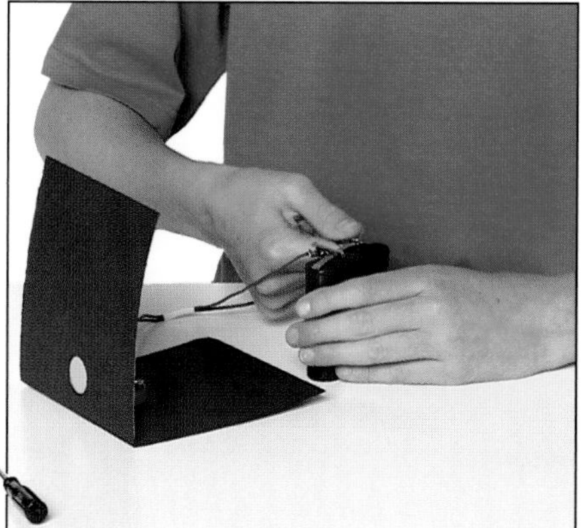

2 COVER EACH HOLE with a piece of tracing paper. Behind each hole, place a flashlight bulb in its holder. Then use wires and clips to connect the bulbs to their batteries.

3 ASK A FRIEND TO MOVE EACH OF THE BULBS closer to or farther away from the holes while you look on from the other side, until all three bulbs appear to be equally bright. The bulb next to the small hole should be quite close, and the bulb next to the largest hole should be the farthest away.

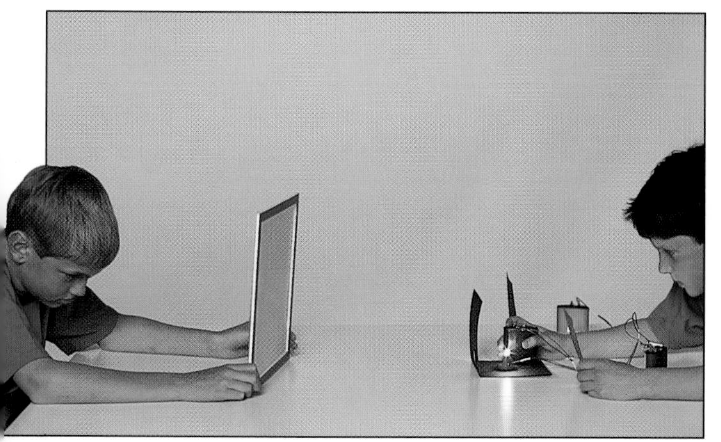

4 GO TO THE FAR END of the darkened room, and hold a piece of tracing paper with a frame in front of you, so that the illuminated holes are all blurred, and you cannot see their sizes directly. Ask your friend to swap the order of the holes.

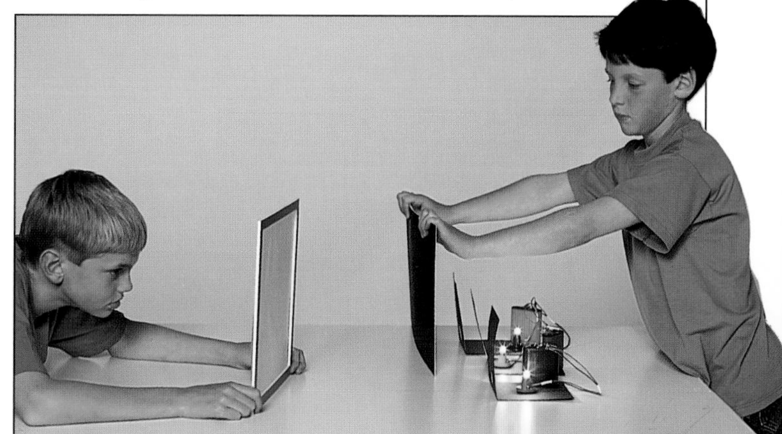

5 ASK YOUR FRIEND to move a black poster-board screen slowly sideways in front of the holes. As each hole appears, count how long it takes to reach its total brightness. Can you tell the holes apart? Like quasars, the smallest hole brightens fastest.

The expanding Universe

TWO MAJOR BREAKTHROUGHS made in this century tell us a great deal about our Universe's birth. In 1929 Edwin Hubble found that the Universe is expanding, which suggests that it had a definite—and violent—beginning in time: the Big Bang. The weak background of radio waves bathing all of space, found in 1965, is probably the cooled-down radiation from this awesome event. These discoveries seem to show that the Universe was born about 15 billion years ago. There was no "before the Big Bang"—space and time were born together. Scientists believe that all matter was created during the Big Bang: the raw material of stars, planets, and galaxies. For a billion years the Universe was too hot to make stars. Then the first galaxies began to form. The Universe keeps expanding, pushing the galaxies apart.

DISCOVERY

Edwin Hubble

American astronomer and cosmologist Edwin Hubble (1889–1953) was the first person to prove that certain "nebulae" were actually galaxies beyond our own. This was a major achievement, which led astronomers to look at the Universe in a new way. Hubble also classified galaxies into different types, and he later found that most galaxies are moving away from us. This led to the realization that the Universe is constantly expanding.

The Doppler effect

Edwin Hubble discovered that nearly all the galaxies are moving away from us, propelled by the expansion of the Universe. The principle that led Hubble to his discovery is one that we encounter in our everyday lives. It is called the Doppler effect, named after the 19th-century Austrian physicist Christian Doppler. The Doppler effect is employed in police radars to catch speeders. Here, you can experiment with the Doppler effect on sound waves.

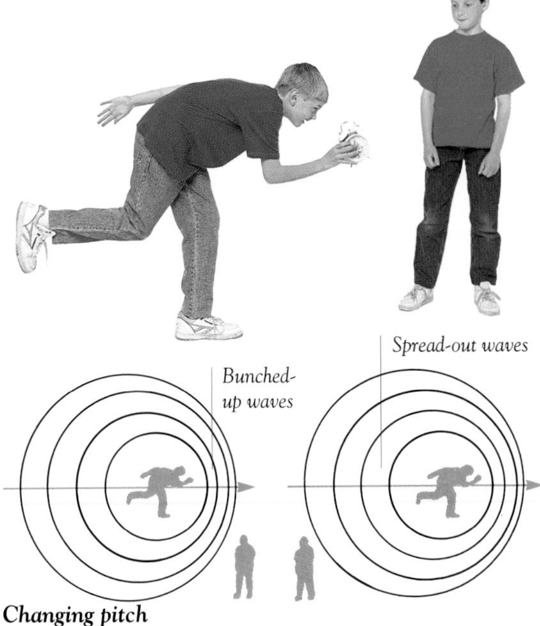

Changing pitch
If a friend runs at you carrying an alarm clock, the sound waves of the alarm will bunch up, creating a high pitch. As the clock passes the waves spread out, and the pitch lowers.

EXPERIMENT

The red shift

When an object emitting light waves moves toward or away from us, its color balance changes because of the Doppler effect. Light from an approaching star or galaxy moves toward short blue wavelengths—showing a "blue shift." Receding stars and galaxies show a "red shift." The bigger the shift, the higher the speed of the object. Hubble discovered the rate at which the Universe expands by measuring the red shifts of the galaxies. He compared the spectrum of a distant galaxy with a laboratory spectrum of a light source at rest. Measuring the amount of shift reveals just how fast the object is moving.

YOU WILL NEED
- *paper* • *compass*
- *black thread*
- *pencil* • *watercolors*
- *ruler* • *glue* • *scissors*
- *poster board*
- *paintbrush*
- *bowl of water*

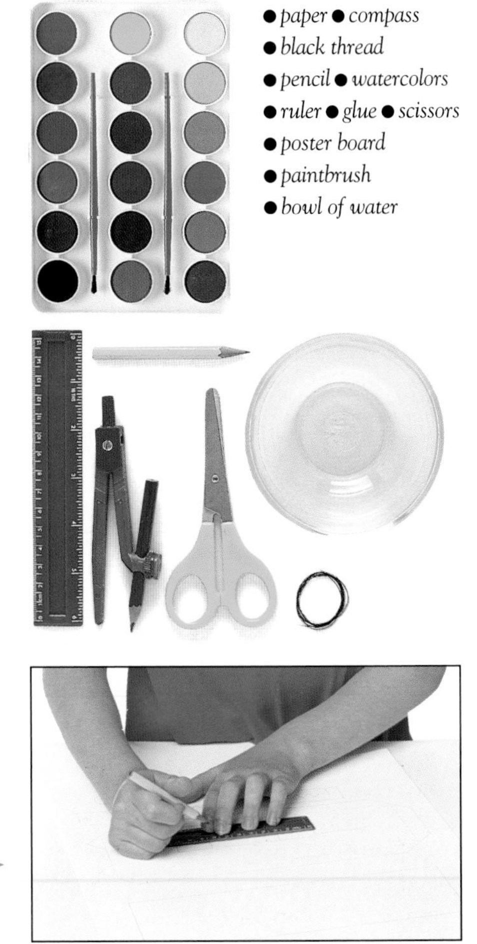

1 COPY THE DIAGRAMS on the opposite page, and try to get the measurements exactly correct. The dotted lines are folds; the solid blue lines are cut marks.

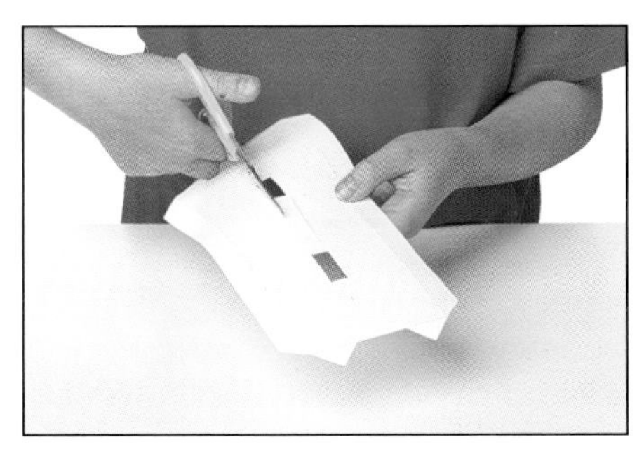

2 CUT OUT THE COPIES of the diagrams. In this photograph the tab in the center of the main template is being cut around. Do not cut it off or the model will not work correctly.

3 PAINT THE MAIN template and the insert as shown below. The template is black with an "Earth" at one end of the hole and a spectrum in the center; the insert is black with white lines.

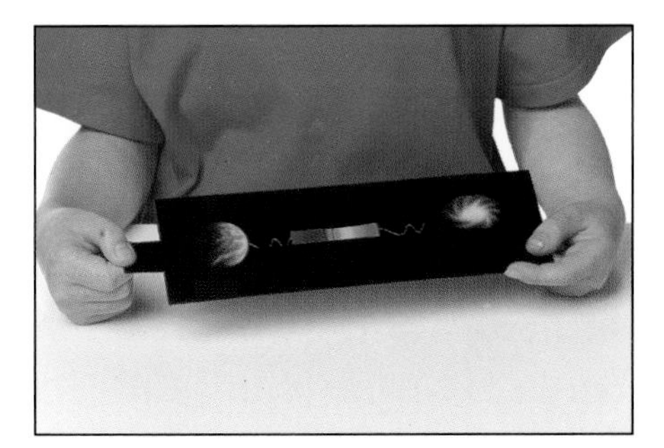

4 PAINT THE "GALAXY" attachment black and white. Once the template, insert, and "galaxy" have dried, you can begin to assemble the model. First, place the insert behind the main template.

The main template
Use the diagram at right as a guide for the measurements for the main body of the model.

The insert
This diagram (right) shows the paper insert for the main template.

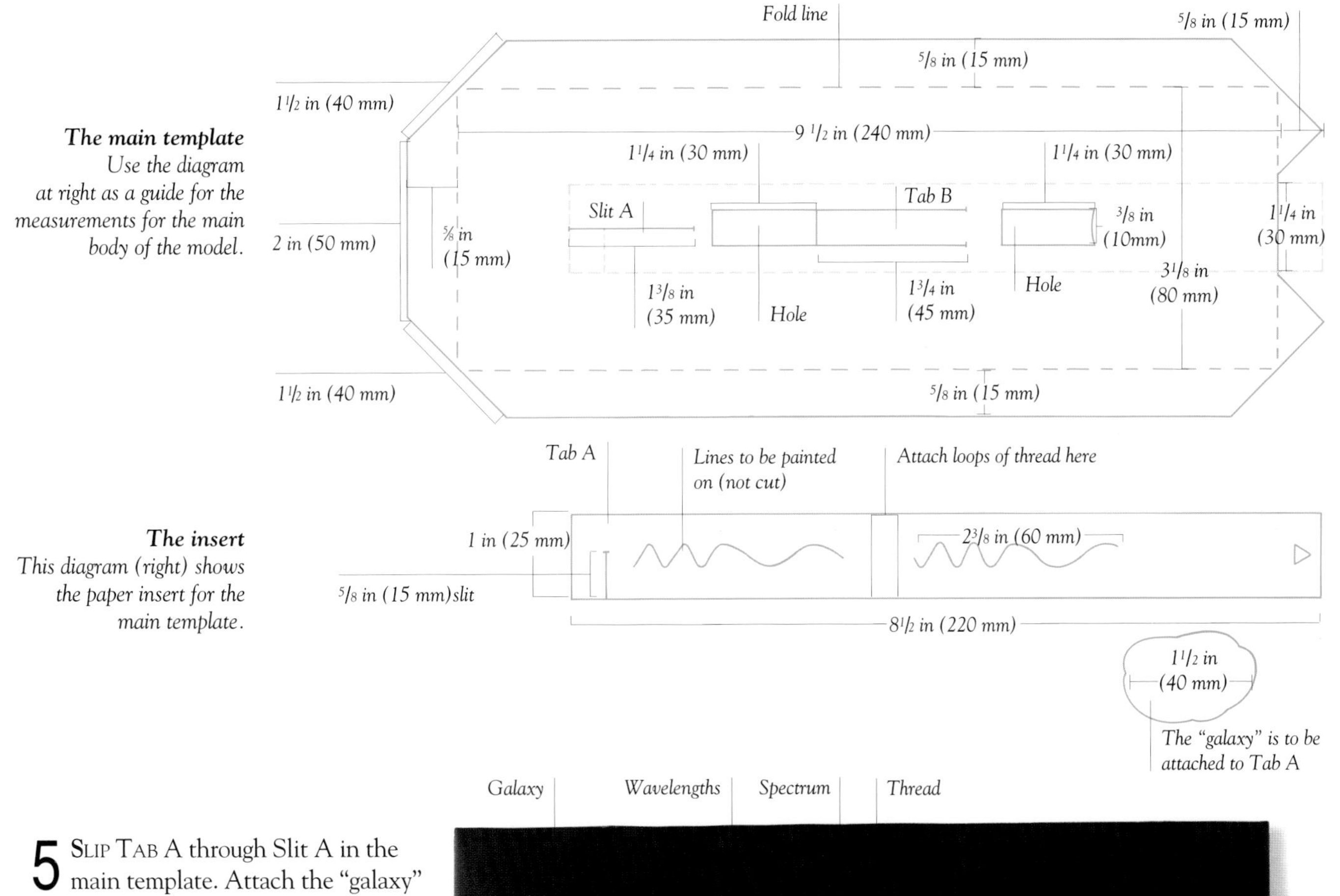

5 SLIP TAB A through Slit A in the main template. Attach the "galaxy" to Tab A with some glue. Then tie two small loops of black thread onto the spectrum as shown at right. Fix the loops firmly to the back of the insert, and slip them over Tab B on the main template. When the loops are in position, take a piece of poster board measuring 8½ x 3⅛ in (220 x 80 mm) and place it behind the main template. Fold the template flaps over the poster board, and glue them down.

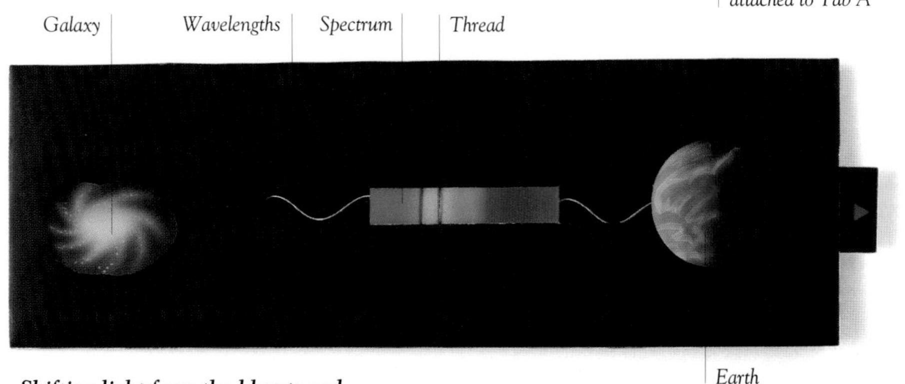

Shifting light from the blue to red
When you pull the insert, the "galaxy" moves toward "Earth." The white lines, showing wavelength, will bunch up, and the black threads (the light indicator) will be in the blue end of the spectrum. Change direction, and the wavelength will widen and the threads move to the red end of the spectrum.

Big Bang to Big Crunch

LOOKING INTO THE UNIVERSE'S PAST is relatively easy. Because of the time it takes light to travel from place to place, we see everything as it was in the past—the farther away, the farther back in time. But we cannot look into the future and see the Universe's fate. So will the Universe continue to expand forever? It all depends on how much matter it contains. The gravity of all the matter in the Universe acts like a brake on the expansion. If there is not enough matter, the Universe will expand forever, growing steadily colder and emptier. If there is enough matter, the expansion will reverse, and the Universe will start to contract, ending in a "Big Crunch." This could eventually lead to a new Big Bang. There is now evidence that 99 percent of the Universe consists of invisible "dark matter," but probably not enough to cause a Big Crunch.

EXPERIMENT

The amazing expanding Universe

Astronomers agree that the Universe began with the Big Bang and that the Universe is now expanding, carrying all the galaxies with it as it grows bigger and bigger. In this experiment you can make a model of the expanding Universe using a balloon.

YOU WILL NEED

- *balloon*
- *felt-tip pen*

1 MARK SEVERAL evenly spaced dots on the balloon. The dots represent galaxies in the expanding Universe.

2 NOW BLOW UP the balloon. Imagine you were situated on a dot. In what direction from you would the other dots be moving?

EXPERIMENT

Dialing the Universe

Timescales in the Universe are incredibly long, but you can get some idea of them by making "Universe dials." Make three dials: one covering the evolution of the Universe from the Big Bang to the present, one showing a cycle of Big Bang-Big Crunch-Big Bang, and the third illustrating an ever-expanding Universe. Predicting the Universe's future requires careful measurement of its past. By comparing its past and present expansion rates, astronomers hope to discover what lies in store. The measurements must be very exact.

YOU WILL NEED

- *paper* • *paints* • *ruler* • *compass* • *glue* • *scissors*
- *glass bowl* • *paintbrushes* • *poster board*

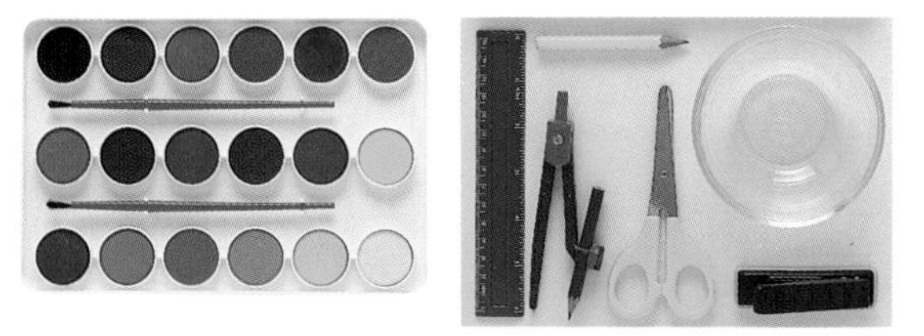

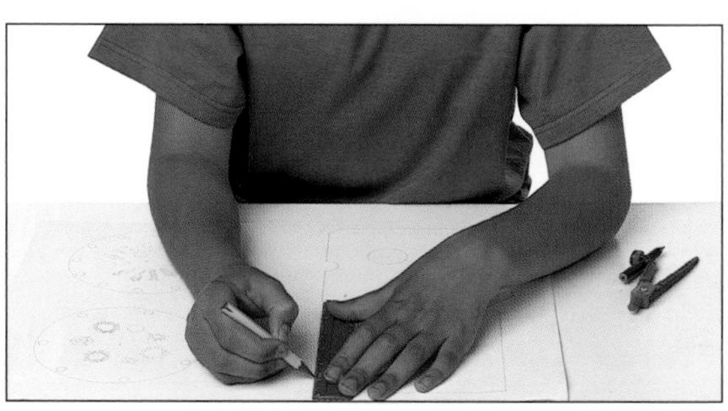

1 DRAW THREE PAPER CIRCLES 5 in (125 mm) in diameter. Draw a 6 x $6\frac{1}{2}$ in (150 x 160 mm) paper sleeve, and trim it as shown in the top left diagram, opposite. Then draw a square on poster board measuring 5 x 5 in (125 x 125 mm).

2 CUT OUT THE FIVE SHAPES. Where there are solid lines on the dials, make a cut with your scissors. On the sleeve, cut around the outside edge and the small circle, draw the triangles, and cut along the solid line at the top and sides of the triangles.

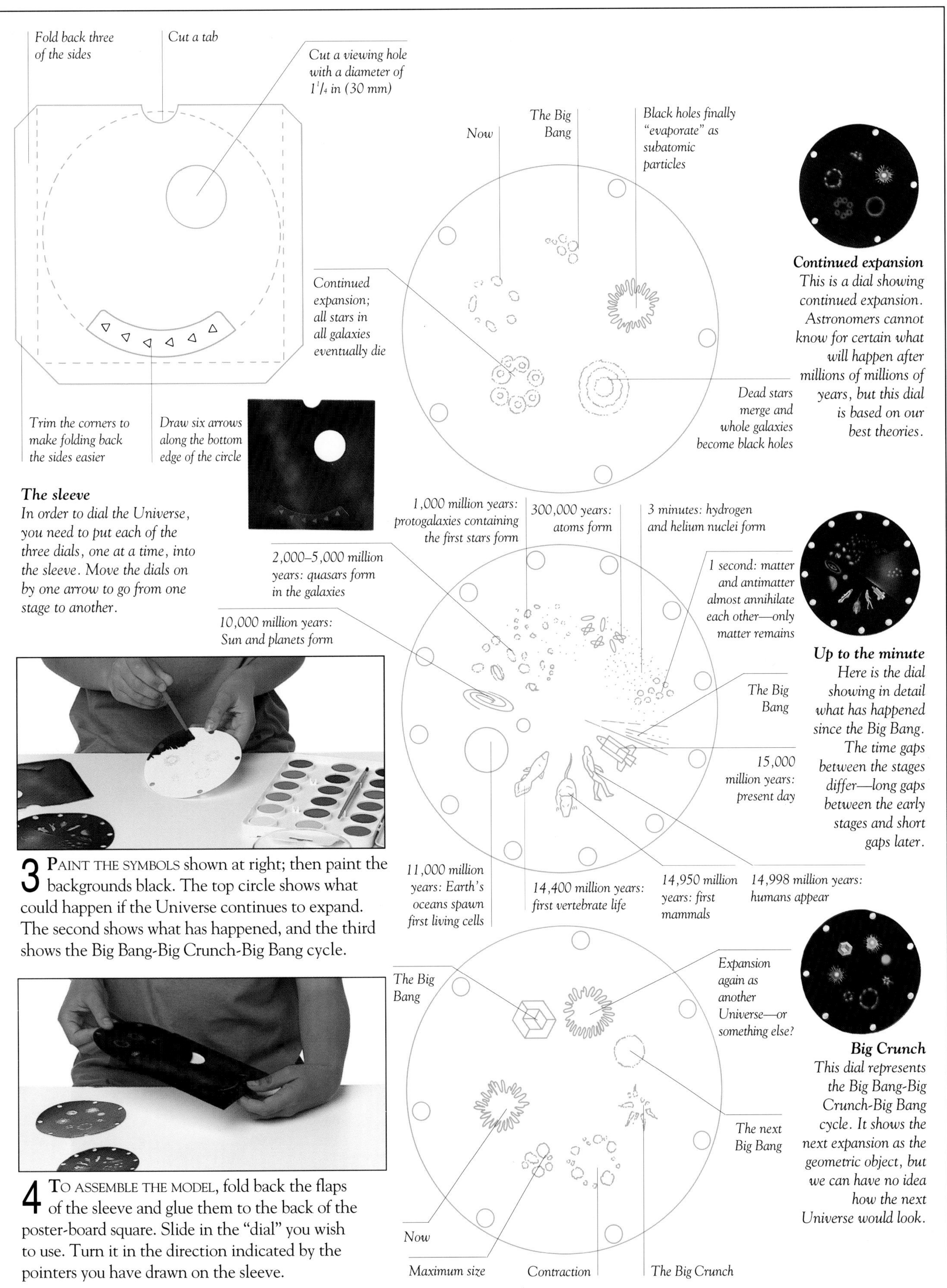

The sleeve
In order to dial the Universe, you need to put each of the three dials, one at a time, into the sleeve. Move the dials on by one arrow to go from one stage to another.

Continued expansion
This is a dial showing continued expansion. Astronomers cannot know for certain what will happen after millions of millions of years, but this dial is based on our best theories.

Up to the minute
Here is the dial showing in detail what has happened since the Big Bang. The time gaps between the stages differ—long gaps between the early stages and short gaps later.

Big Crunch
This dial represents the Big Bang-Big Crunch-Big Bang cycle. It shows the next expansion as the geometric object, but we can have no idea how the next Universe would look.

3 PAINT THE SYMBOLS shown at right; then paint the backgrounds black. The top circle shows what could happen if the Universe continues to expand. The second shows what has happened, and the third shows the Big Bang-Big Crunch-Big Bang cycle.

4 TO ASSEMBLE THE MODEL, fold back the flaps of the sleeve and glue them to the back of the poster-board square. Slide in the "dial" you wish to use. Turn it in the direction indicated by the pointers you have drawn on the sleeve.

Is anyone there?

THE EARTH IS THE ONLY planet we know where life has developed. Yet our Sun is a very ordinary star, and astronomers believe that many of the other 200 billion stars in our Galaxy must have planets like ours. Would life develop there as well? If other life forms are anything like the humans of planet Earth, they will want to communicate with each other across the vastness of space. At the moment astronomers interested in SETI —the Search for Extra-Terrestrial Intelligence—are trying to pick up radio signals from alien civilizations. The problem is that we don't know what messages the aliens might be sending or what frequencies they might be broadcasting on. NASA has begun a "listening-in" project that is the equivalent of tuning in to millions of cosmic radio stations at once.

EXPERIMENT

Decipher an alien message

Instead of sending radio messages in a spoken language, aliens would probably use some sort of mathematical code. Scientists on the Earth would then have to decipher the message. Here is one way to send and decipher an interstellar picture. Use the following numbers as a code: 10111010 01110000111000001000111 1110011100001110000111000 0010100001010010101 01.

YOU WILL NEED

- *paper* • *pencil* • *black pen*
- *ruler* • *writing pad*

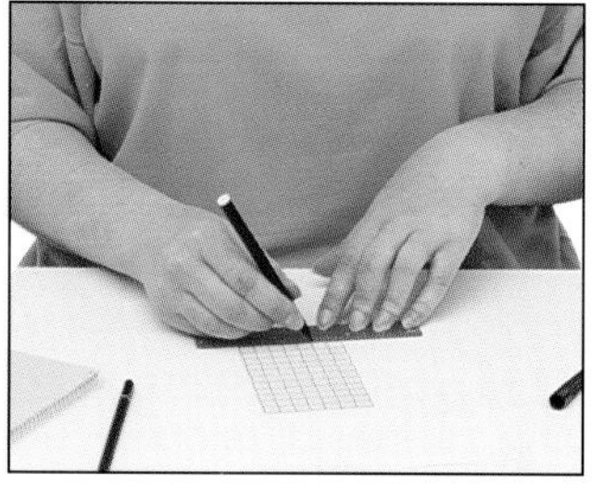

1 COUNT THE digits (every 1 and 0). Work out what two numbers, multiplied, give this total. Draw a grid of ½ x ½ in (1 x 1 cm) boxes. The large number tells you how many boxes tall to make the grid; the smaller, its width.

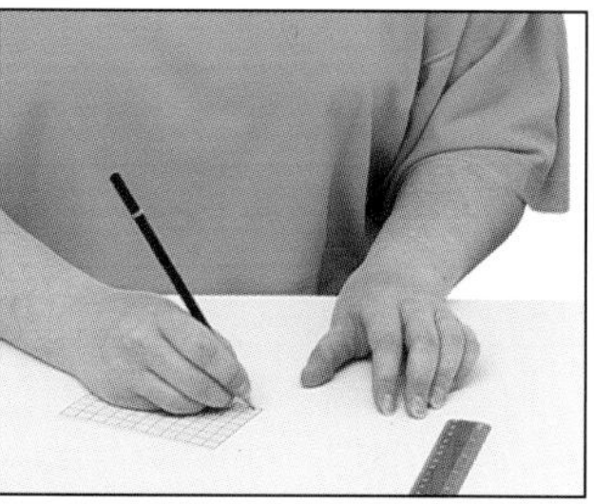

2 WRITE IN each 1 and 0 in the code, starting at the top left-hand corner and going from left to right across the rows. Color in each box with a 1 inside. What is the message? Now make your own message.

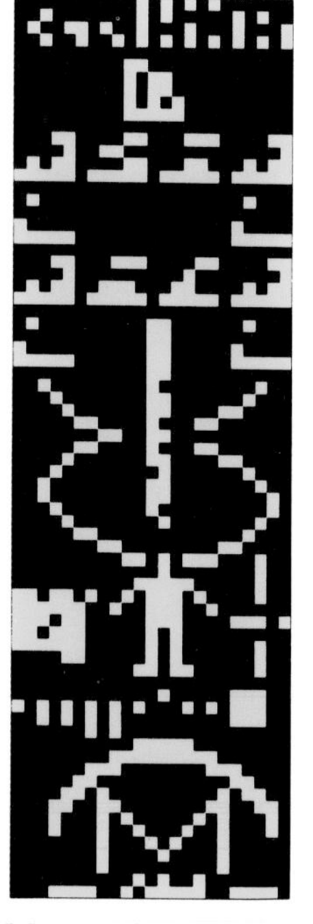

Message from Earth
Frank Drake beamed this message into space from a radio telescope to tell aliens about Earth life. You can make your own message. Make a grid of boxes with the sides equal to two prime numbers (p.154). Color in the squares to make a picture. For the code, write a 1 for each colored square and a 0 for each white square.

EXPERIMENT

How many aliens are there?

An American astronomer named Frank Drake worked out an equation that can tell us how many alien civilizations in the Galaxy are trying to make contact with us. The answer comes after you have put in a series of likely numbers that you can estimate for yourself. The device described here is an alien calculator based on Drake's equation. Try it out yourself, and then ask your friends to try it and see if they get a different answer from yours.

YOU WILL NEED

- *poster board* • *black pen* • *pencil* • *ruler*
- *scissors* • *graph paper* • *six ¾-in (16-mm) strips of colored poster board*

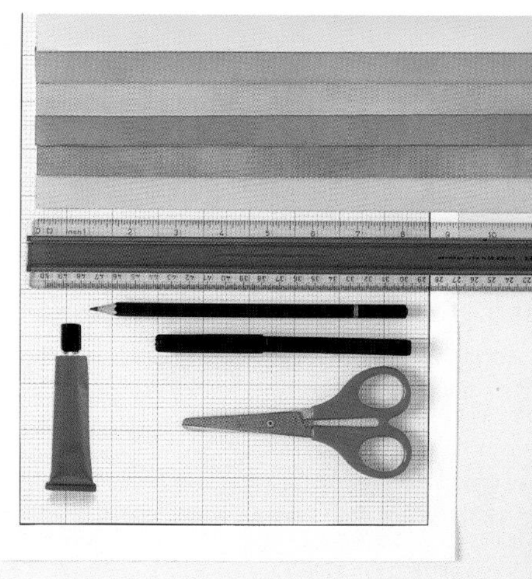

What would aliens look like?
An alien's appearance would depend on its planet —high-gravity planets would have squat aliens. But aliens would probably have "eyes" sensitive to the light of their star and "ears" for hearing.

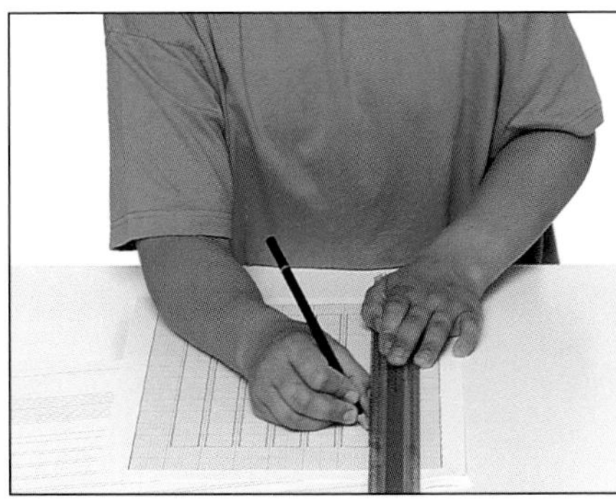

1 **MARK A** 1 x 1 in (2 x 2 cm) grid on the graph paper, 9 boxes deep and 15 boxes wide. Then draw six long boxes in the positions shown below, and cut them out, leaving six holes.

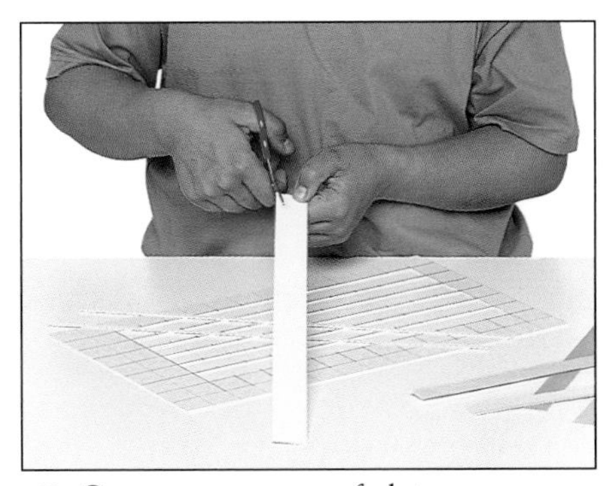

2 **CUT SEVEN STRIPS** of plain poster board the same width as the spaces in between the holes. Flip the graph paper. Glue the strips on the spaces and above and below the top and bottom spaces.

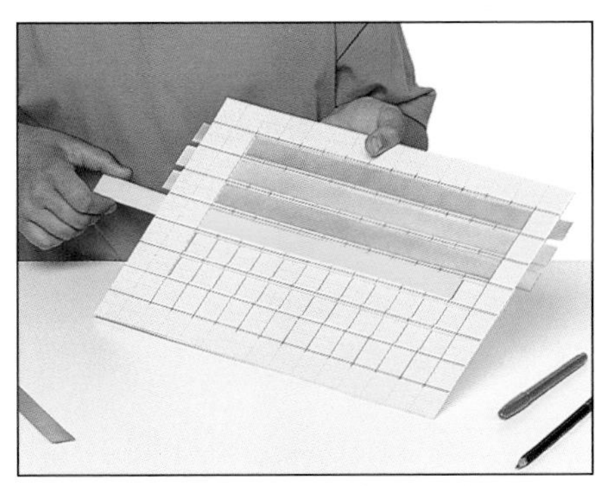

3 **COVER THE BACK** of the graph paper with poster board, and glue it at the top and bottom. Slide in the colored strips so that they align at the right. Then add all the text shown below, in position.

Astronomers know this figure, so circle it and use it as your starting point

Number of stars in the Galaxy — 200,000 million

The small arrows help you to line up the values

Fraction of stars with planets (F_P) — 100% 10% 1% 0.1%

Number of Earth-like planets (N_E) — 10 1 1 in 10 1 in 100

Fraction where life starts (F_L) — 100% 10% 1% 0.1%

Fraction with intelligence (F_I) — 100% 10% 1% 0.1%

Lifetime of a civilization (L) — 10,000 million 1,000 million 100 million 10 million 1 million 100,000 10,000 1,000 100 1 Years

This arrow will show the number of alien civilizations

Number of civilizations (N) — 100,000 million 10,000 million 1,000 million 100 million 10 million 1 million 100,000 10,000 1,000 100 10 1

Counting your aliens
As you choose a value for each category, circle it in pencil. Then slide the next strip along so that its arrow lines up with this circle. As you proceed, the strips will form a step pattern.

4 **F_P IS THE FRACTION** of stars with planets. Line up the top arrow with the circle above. Estimate a planet value, and circle it on this strip in soft pencil.

5 **N_E IS THE NUMBER** of planets in each planetary system that are like the Earth; they need liquid water. Line up the arrow with your F_P, and choose N_E.

6 **F_L IS THE FRACTION** of Earth-like planets on which life has started. Line up the arrow with your N_E, and choose F_L—if life is rare, perhaps 1%.

7 **F_I IS THE FRACTION** of life-bearing planets where intelligent life—like human beings—has evolved. Line up the arrow, and decide on your value for F_I.

8 **L IS THE LIFETIME** of a civilization. Line up the arrow, and circle your choice. If a civilization lasts only a few years, would any exist now?

9 **N IS THE ANSWER** you want: the number of civilizations in the Milky Way. Follow the arrow to the main scale, and read your answer.

Stars of the northern skies

THIS CHART SHOWS all the stars visible from northern latitudes on the Earth. You can get an idea of the stars that are visible tonight by rotating the book so that the current month is at the bottom—the stars in the lower region lie to the south in your sky. To find out more accurately, use this chart to make a planisphere (pp.112–113).

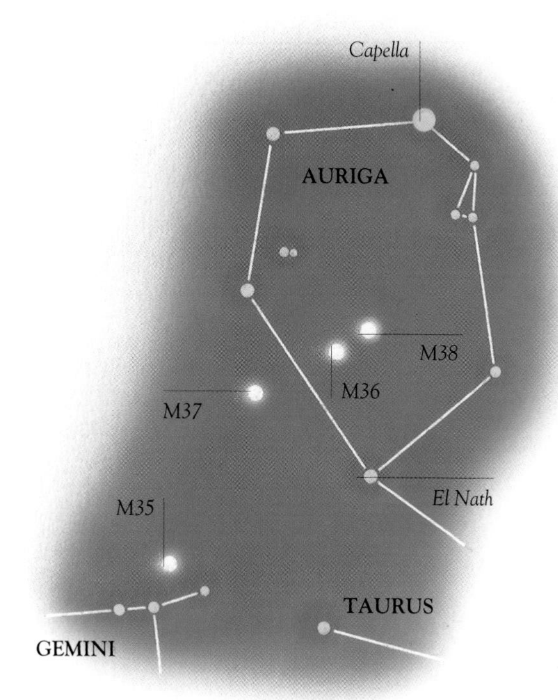

Winter clusters
With binoculars, you can make out some fine star clusters in the winter sky in the region of Auriga and Gemini: M35 is just visible to the naked eye. These clusters lie about 3,000 light-years away.

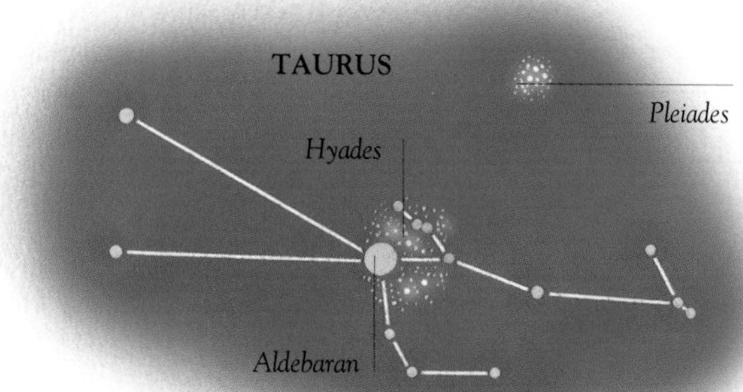

Nearby star groups
A gorgeous sight in fall and winter is the Pleiades cluster of stars. It is called the Seven Sisters, but with good eyesight you can observe more than seven stars. How many can you see? Below in this picture is the scattered Hyades, the nearest star cluster to the Sun.

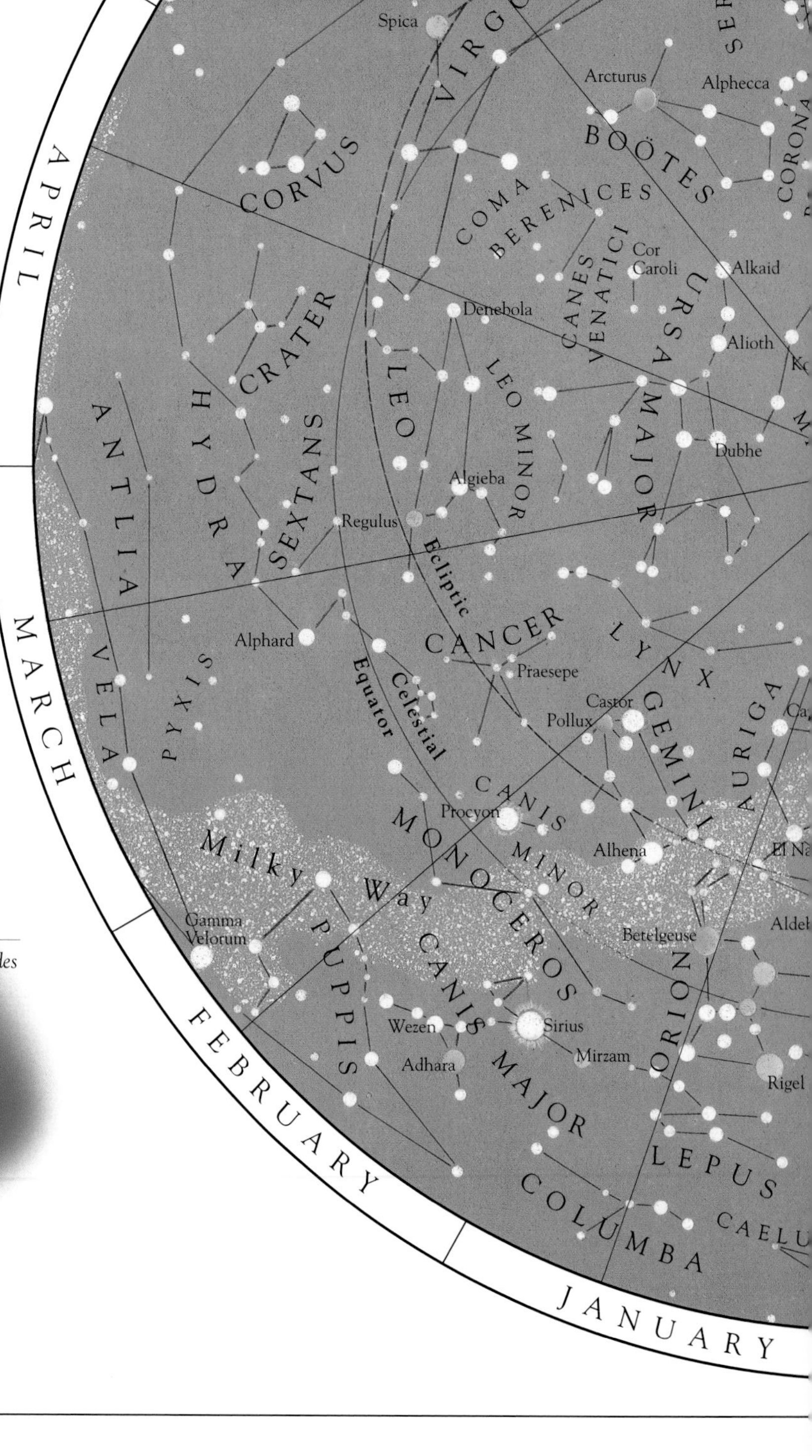

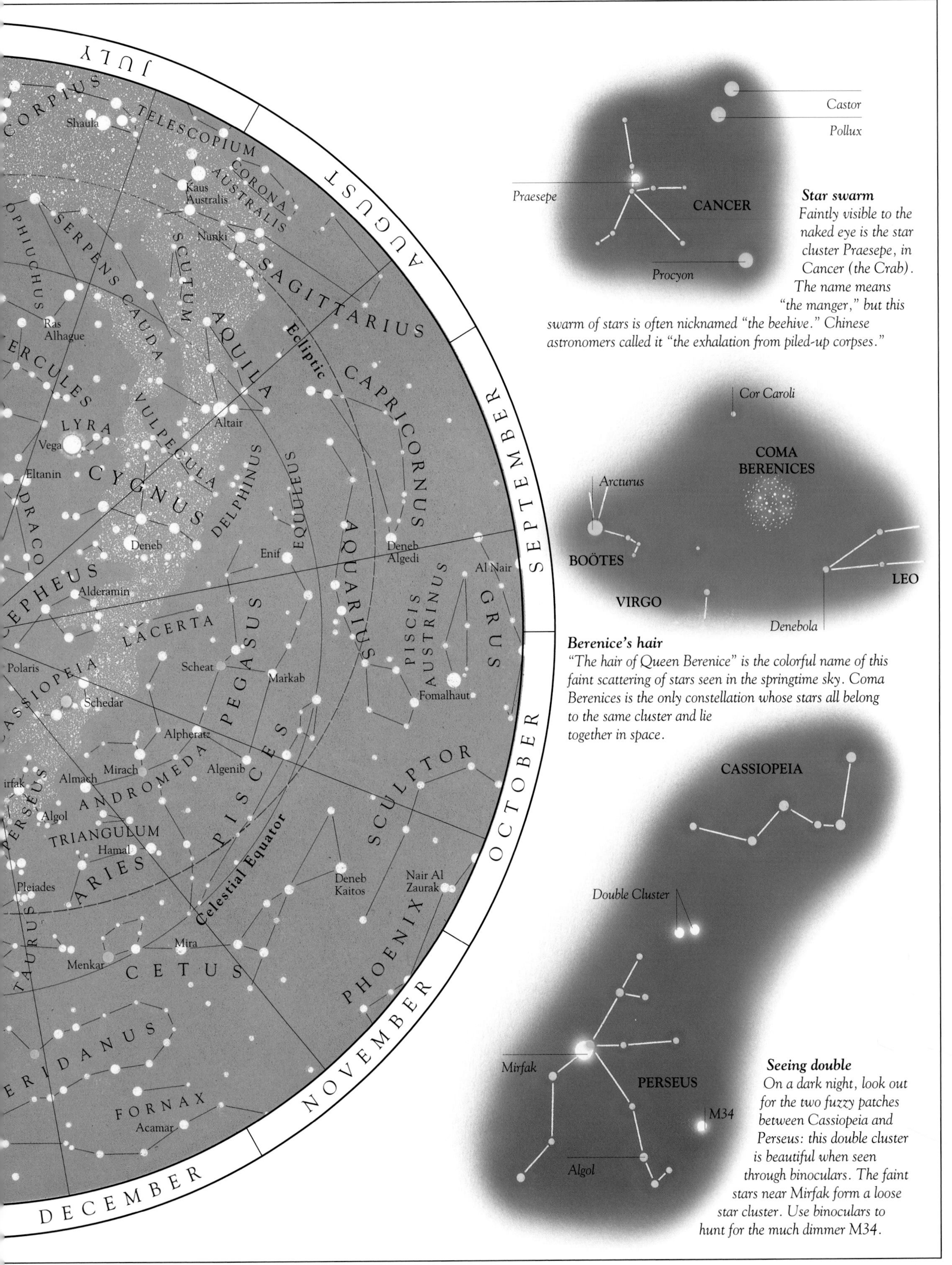

Star swarm
Faintly visible to the naked eye is the star cluster Praesepe, in Cancer (the Crab). The name means "the manger," but this swarm of stars is often nicknamed "the beehive." Chinese astronomers called it "the exhalation from piled-up corpses."

Berenice's hair
"The hair of Queen Berenice" is the colorful name of this faint scattering of stars seen in the springtime sky. Coma Berenices is the only constellation whose stars all belong to the same cluster and lie together in space.

Seeing double
On a dark night, look out for the two fuzzy patches between Cassiopeia and Perseus: this double cluster is beautiful when seen through binoculars. The faint stars near Mirfak form a loose star cluster. Use binoculars to hunt for the much dimmer M34.

Stars of the southern skies

ON THIS CHART are all the stars that can be seen from the Southern Hemisphere of the Earth. To get an idea of which stars are visible tonight, face north and turn the book until the current month is at the bottom. The sky in front of you should match the stars in the lower part of the chart. Use this chart to make a planisphere (pp.112–113) to obtain a more accurate guide to the sky. Start by identifying the brighter stars, then the fainter stars, and finally, interesting objects such as the star clusters that are shown to the side of the main map.

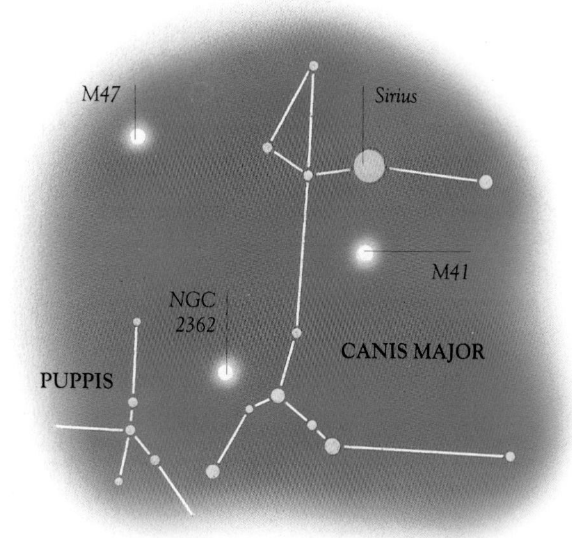

Dog stars

Several star clusters lie near Sirius, the brightest star in the sky. On a clear night you can see M47 with the naked eye as a fuzzy patch. Use binoculars to hunt for M41 and NGC 2362.

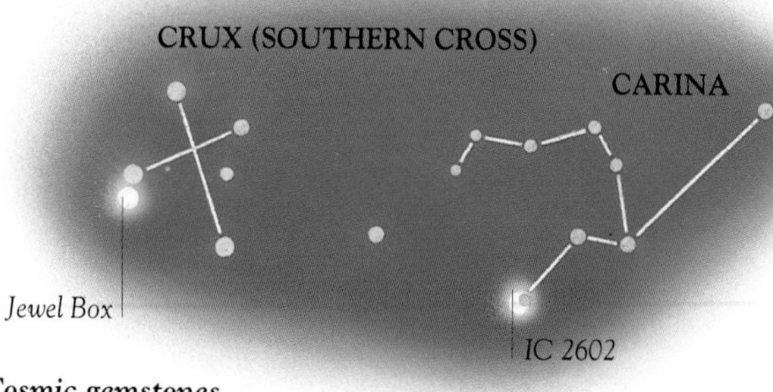

Cosmic gemstones

The Southern Cross is your guide to two lovely clusters. Through a telescope, the Jewel Box looks like a heap of blue and red gems. IC 2602 is easily seen with the naked eye; it rivals the more famous Pleiades (the Seven Sisters).

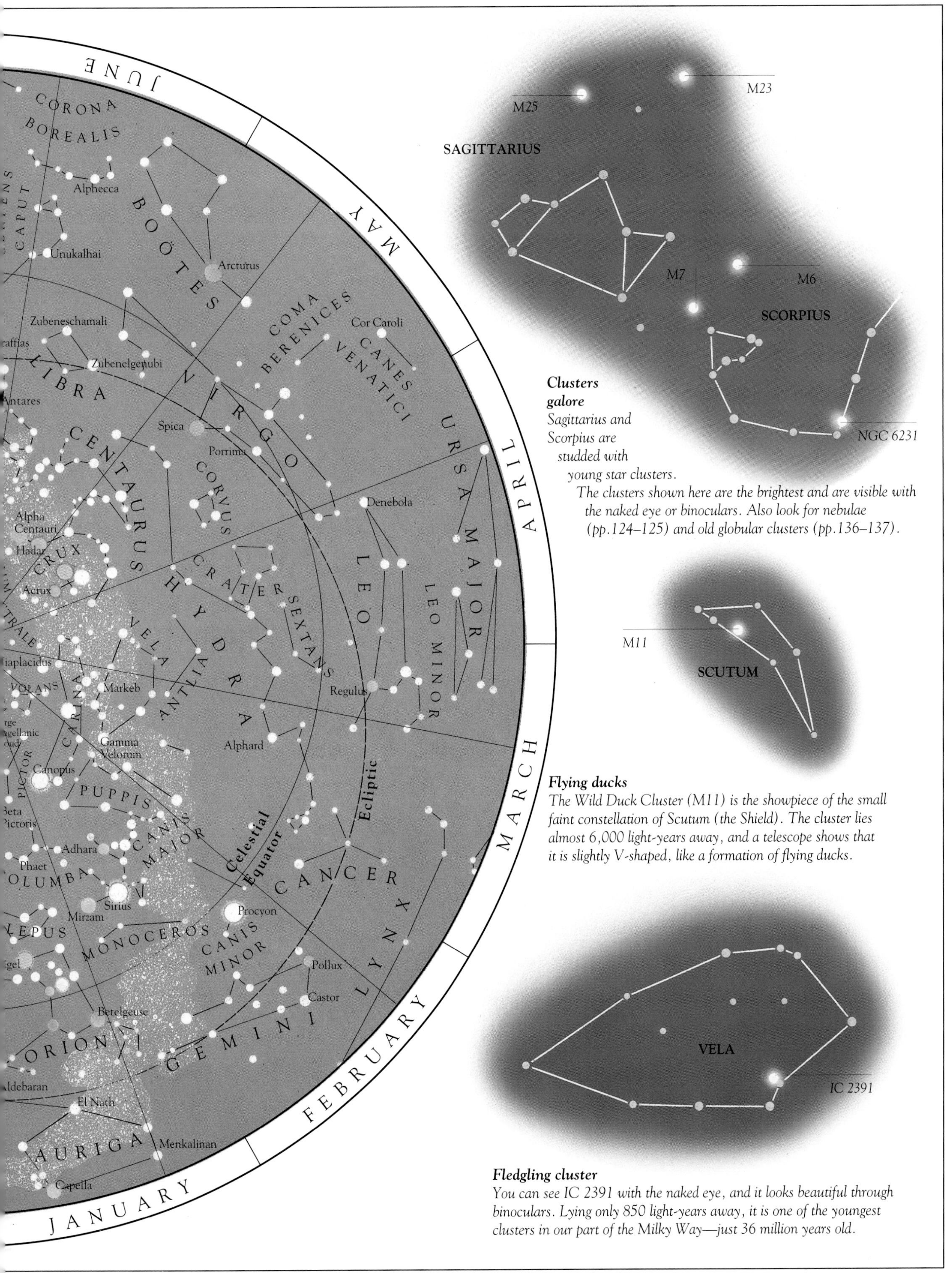

Clusters galore
Sagittarius and Scorpius are studded with young star clusters. The clusters shown here are the brightest and are visible with the naked eye or binoculars. Also look for nebulae (pp.124–125) and old globular clusters (pp.136–137).

Flying ducks
The Wild Duck Cluster (M11) is the showpiece of the small faint constellation of Scutum (the Shield). The cluster lies almost 6,000 light-years away, and a telescope shows that it is slightly V-shaped, like a formation of flying ducks.

Fledgling cluster
You can see IC 2391 with the naked eye, and it looks beautiful through binoculars. Lying only 850 light-years away, it is one of the youngest clusters in our part of the Milky Way—just 36 million years old.

Glossary

ABSORPTION LINE A narrow zone in the absorption spectrum of a gas where light is absorbed, also known as a Fraunhofer line. The absorption lines of a chemical element are its own unique "fingerprint": they tell astronomers what substances make up the Sun and stars.

ACCELERATION The rate at which velocity changes with time, caused by the application of a force. The force of gravity gives rise to the "acceleration due to gravity"—the acceleration felt by a body falling in a gravitational field, such as that of a planet. On the Earth this acceleration is 32 ft (9.8 m) per second per second.

ACCRETION DISK Disk of hot, glowing matter spiraling into a black hole.

ALTITUDE The angular elevation of a celestial object above the horizon.

ASTEROID One of thousands of chunks of rock and metal, whose orbits are generally confined to a belt between Mars and Jupiter.

ASTROLABE An ancient instrument, dating from the second century B.C., which shows the positions of the Sun and the stars.

ATMOSPHERE A layer of gas surrounding a planet; the outer parts of a star.

ATOM The smallest unit of an element—such as hydrogen or oxygen.

AURORA The green and red glow that is sometimes visible in the sky near the polar regions, caused by energetic particles from the Sun (possibly from a flare) exciting the gases in the Earth's atmosphere.

AXIS An imaginary line that passes through the poles of a body, such as the Earth, about which it rotates.

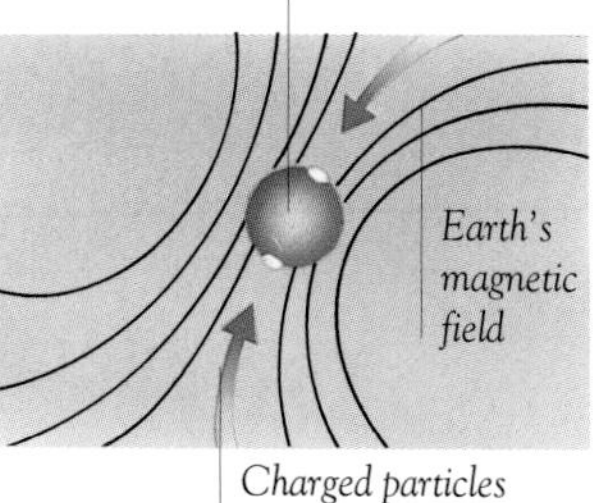

The auroras
The Earth's magnetic field channels charged particles from the Sun toward the poles, where they cause the aurora australis (Southern Hemisphere) and the aurora borealis (Northern Hemisphere).

BARYCENTER The center of gravity of a pair or a group of bodies, about which they all revolve.

BIG BANG In theory, a violent event that gave birth to the Universe some 15 billion years ago.

BIG CRUNCH The ultimate collapse of the Universe that may take place in the future if the Universe contracts.

BINARY SYSTEM A system of two stars that are in orbit around each other.

BLACK HOLE A collapsed object whose gravity is so strong that nothing—not even light—can escape it. As a result, the object is black, and it is a hole because things that fall "in" can never escape its gravity.

BROWN DWARF An object that is smaller than a star but larger than a planet. It has no nuclear reactions but produces infrared radiation because it is gradually shrinking. This compression makes the brown dwarf warm.

CATADIOPTRIC TELESCOPE A telescope that gathers light with a combination of lenses and mirrors. A "cat" is a compact telescope with a wide field of view.

CELESTIAL SPHERE The imaginary bowl of sky that surrounds the Earth. Astronomers measure star positions according to the stars' latitude or longitude on the celestial sphere.

CENTRIFUGAL FORCE An apparent outward force felt when moving in a circle—for example, at the surface of a spinning body or in an orbit. The "force" is in fact the natural tendency of a body to move in a straight line rather than a circle.

CEPHEID STARS Very luminous variable stars whose periods of variability are linked to their actual brightness. Astronomers use Cepheid stars to measure distances in space.

CHROMOSPHERE The lower layer of the Sun's atmosphere. It shines pinkish red.

CLUSTER OF GALAXIES A group of galaxies held together by its own gravity.

COMET A body made of ice and dust that can develop a huge gaseous head and long tail when it approaches (and is warmed by) the Sun.

CONSTELLATION An imaginary group of stars in the sky, which has usually been given a mythological name (like Leo or Orion). Stars in a constellation are not usually associated, but spread throughout space.

CORE (OF A PLANET) The innermost part of a planet, which is made of rock or metal (which may be liquid) under considerable pressure. A fast-spinning metal core can produce a magnetic field encompassing the planet.

CORONA The Sun's outer atmosphere, which is visible as a pearly halo during a total solar eclipse.

COSMIC RAY A tiny, fast-moving electrically charged particle coming from space.

COSMOS Another term for the Universe.

CRATER A saucer-shaped hole that was blasted in the ground by an explosion. The craters on the Moon and planets were formed by the impacts of meteorites.

CRUST The rocky surface layer of a planet.

DARK MATTER Invisible matter that is thought to make up 99 percent of the mass of the Universe.

DEGREE A measure that is 1/360 of a circle. Ninety degrees make a right angle; 180 degrees make a straight line.

DENSITY The degree of "solidity" of a body: its mass divided by its volume.

DOPPLER EFFECT The change in the observed frequency of sound or radiation that takes place when the observer and the source are moving relative to each other.

DUST Microscopic grains in space that absorb starlight. The grains are "soot" from cool stars, and they sometimes clump together in huge dark clouds.

EARTHSHINE A faint illumination of the unlit part of the Moon, caused by sunlight reflecting off the Earth's atmosphere. This phenomenon is sometimes referred to as the "old Moon in the new Moon's arms."

ECLIPSE An effect caused when one celestial body casts a shadow on another. In an eclipse of the Moon, the Moon moves into the Earth's shadow. An eclipse of the Sun takes place when the Moon's shadow falls on the Earth, blocking out the Sun.

ECLIPSING BINARY Two stars in close proximity (a double star) that periodically seem to pass behind and in front of each other when viewed from the Earth.

ELECTRONS Tiny particles with a negative electrical charge that occupy the outer parts of an atom.

ELLIPSE An oval shape. The planets move around the Sun in elliptical orbits.

ELLIPTICAL GALAXY A galaxy that has an oval or round shape, with no spiral arms. Elliptical galaxies contain very little dust and gas, and they are made up mostly of old red stars.

EMISSION LINE A narrow, bright zone in the emission spectrum of a hot gas.

EQUINOX The two days of the year—roughly March 21 and September 23—when day and night are of the same length. They correspond to times when the Earth is tilted on its axis so that the Sun is overhead at the Equator.

ESCAPE VELOCITY The speed at which a launch vehicle needs to travel in order to escape the surface gravity of any celestial body. Escape velocity depends on size as well as mass: the smaller the object, the higher the escape velocity.

EYEPIECE A small lens that is placed at the viewing end of a telescope to magnify the image that is produced by the mirror or main lens.

FILAMENTS AND VOIDS Strings of superclusters surrounding empty regions of space—the structure of the Universe on the largest scale.

FLYBY The flight of a space probe past a planet, comet, or asteroid, without stopping to orbit or land on it.

FOCAL LENGTH The distance between a lens or mirror, and the point at which the light rays gathered are brought to a focus.

FOCUS The point in a telescope where light rays that have been gathered by the main lens or mirror come together; the place where the image of the object being viewed is formed.

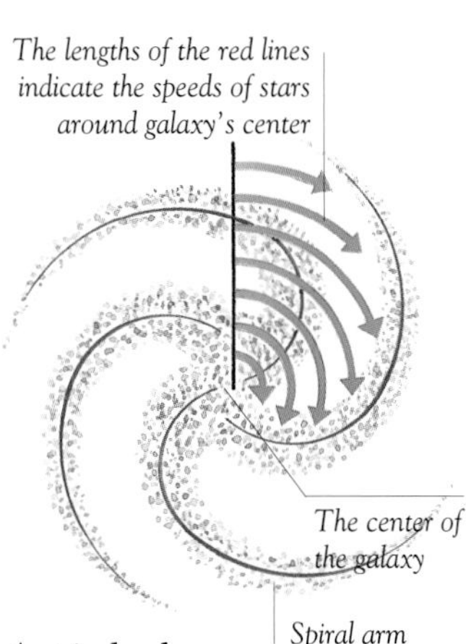

A spiral galaxy
Stars in a spiral galaxy revolve around the center at different speeds. Astronomers believe that gravity helps to keep the spiral arms from winding tightly into the center.

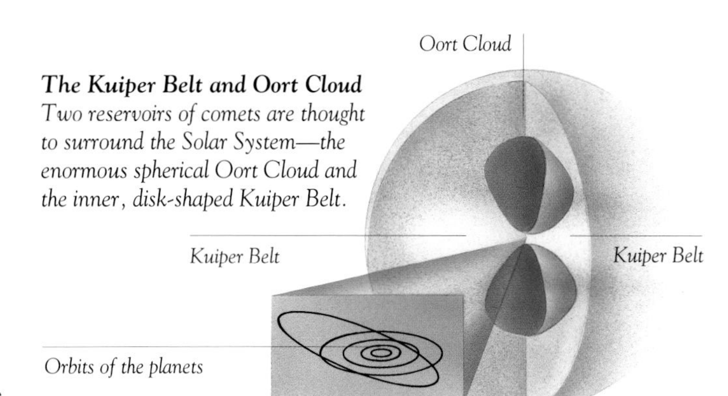

The Kuiper Belt and Oort Cloud
Two reservoirs of comets are thought to surround the Solar System—the enormous spherical Oort Cloud and the inner, disk-shaped Kuiper Belt.

FREQUENCY The number of vibrations that a wave (of sound or radiation) makes in a single second.

FRICTION "Rubbing" between bodies or substances that causes heat to develop. The friction between air and a meteor plunging into the Earth's atmosphere produces so much heat that the meteor burns up.

GALAXY A "city" of millions of stars. There are millions of individual galaxies in the Universe, and they are separated by empty space.

GAS GIANT A big planet, such as Jupiter or Saturn, that is made largely of a very deep, dense gaseous atmosphere.

GRAVITY The force of attraction that is felt between two or more bodies, such as the pull between the Earth and the Moon.

GREENHOUSE EFFECT The rise in temperature caused by the tendency of some gases—such as methane and carbon dioxide—to trap the heat radiated from a planet's surface.

INFLATION An event that in theory occurred fractions of a second after the Big Bang, when the Universe suddenly grew much larger, smoothing out any irregularities.

INFRARED Heat radiation. Its wavelength is between light and radio waves.

IRREGULAR GALAXY A galaxy that has no particular shape. Irregulars are generally small galaxies that contain a mix of old and young stars, and often a great deal of gas to make more stars in the future.

KUIPER BELT A disk-shaped "reservoir" of comets believed to extend from beyond the orbit of Pluto to the inner edge of the Oort Cloud.

LATITUDE The angular distance north or south of the Equator of a spherical body (such as the Earth). Latitude is measured in degrees.

LIGHT POLLUTION The combination of bad street lighting and atmospheric pollution that obscures our view of the stars. The upwardly directed light reflects off the particles in the atmosphere to create a bright, ugly haze.

LIGHT-YEAR The distance a ray of light travels in 1 year (6 million million miles [9.5 million million km]).

LONGITUDE The angular distance east or west of an imaginary line (the meridian) on a spherical body, such as the Earth. Longitude is measured in degrees.

LUMINOUS Glowing. A term usually applied to objects that generate their own light, such as the stars.

MAGNETIC FIELD Magnetism generated in the core of a planet (or on the surface of a star) that extends into space.

MAGNETOSPHERE The magnetic bubble that surrounds each planet, in which the planet's magnetic field is kept separate from the solar wind.

MAGNITUDE The brightness of a star or planet, expressed on a scale of numbers. Bright objects have magnitudes in low numbers; dim objects in high numbers.

MANTLE The rocky layer inside a planet, between the crust and core.

MARE (PLURAL: MARIA) A large, dark marking on the Moon, thought by early astronomers to be a body of water ("mare" means "sea" in Latin). The maria are really huge lava-flooded basins.

MASS The amount of matter that makes up a body. On the Earth, the mass of a body is equal to its weight.

METEOR A shooting star—a small grain of rock that burns up when it enters the Earth's atmosphere.

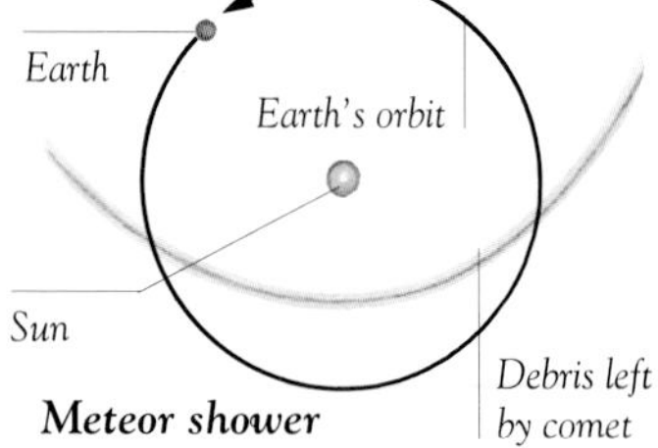

Meteor shower
Meteor showers occur each year when the Earth crosses the debris left by a comet. The comet's dust enters the Earth's atmosphere, and burns up as "shooting stars."

METEORITE A meteor big enough to survive the journey through the atmosphere. When it arrives on a planet's surface, it may form a crater.

METEOR SHOWER A swarm of meteors that is visible when the Earth crosses the orbit of an old comet.

MICROWAVE BACKGROUND The radio background that evenly bathes the Universe. It is believed to be cooled-down radiation from the Big Bang.

MOLECULE A chemical combination of two or more atoms. Dozens of molecules occur naturally in space.

MOON A planet's natural satellite. The Earth's satellite is named "the Moon," but the moons of other planets have unique names, such as Triton.

NEBULA A cloud of gas and dust in space.

NEUTRON An electrically neutral particle in the nucleus of an atom.

NEUTRON STAR A collapsed star that is composed mainly of neutrons.

NOCTURNAL An old instrument that was used for telling the time at night by measuring the position of the Pointers, the last two stars in the "bowl" of the Big Dipper.

NOVA A white dwarf star in a binary system that flares up thousands of times in brightness when matter is dumped on it by another star.

NUCLEAR FUSION A reaction in which the nuclei of light elements, under extreme heat and pressure, combine to make a heavier element. The energy released keeps the stars shining.

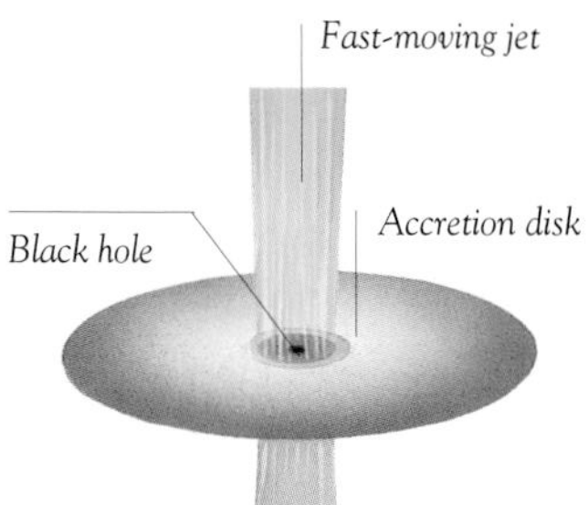

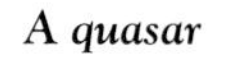

A quasar
The center of a quasar probably consists of a black hole surrounded by a very energetic disk of infalling matter, which can beam jets of fast-moving particles thousands of light-years into space.

NUCLEUS The small, but massive, center of an atom.

OBSERVATORY A place where scientists undertake astronomical research. There are optical observatories, radio observatories, and even observatories in space.

OCCULTATION The passing of one astronomical body in front of another—usually when a large body blocks a smaller body; for example, when the Moon passes in front of a distant star.

OORT CLOUD A "reservoir" of frozen comets that is thought to surround the Solar System in a huge spherical cloud.

ORBIT The path followed by a planet, a satellite, or a star around a more massive body or around a barycenter.

PARALLAX The shift in a nearby object's position when it is seen from two different vantage points. Astronomers use parallax to find the distances to nearby stars.

PENUMBRA The outer, lighter part of a sunspot. Also, the lightest part of a shadow caused by an eclipse.

PHASE The size of the illuminated portion of a planet or moon.

PHOTOSPHERE The visible surface of the Sun or a star.

PLANET A dark, low-mass body in orbit around a star.

PLANETARIUM A "star theater." Stars, planets, and moons are projected onto a darkened dome to simulate the appearance of the night sky.

PLANETARY NEBULA The shell of gas puffed off by an aging star before it becomes a white dwarf.

POLE STAR Polaris, the star that appears to be almost exactly above the Earth's North Pole. It remains in the same position as the Earth spins beneath it.

PRIME NUMBER A number that cannot be divided by any number except itself and 1.

PRISM A block of glass or clear acrylic plastic that disperses a beam of light into a spectrum of colors. Today "diffraction gratings" are more commonly used than prisms to produce the spectra of meteor showers and galaxies.

PROMINENCE A huge arc of gas hanging in the lower parts of the Sun's corona.

PROTON A positively charged particle that forms part of the nucleus of an atom.

PULSAR A young, fast-spinning neutron star.

QUASAR The brilliant and active core of a distant young galaxy, whose outer regions are often too faint to be visible.

RADAR A device that bounces radio waves off a body to calculate its distance or to make maps.

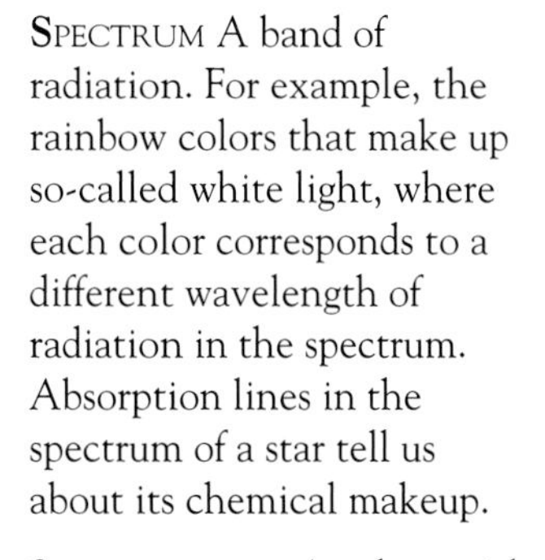

Radiant A point in the sky from which, as a result of perspective, meteor showers appear to originate.

Radio galaxy An active galaxy that gives out as much energy in radio waves as it does in light. Most of the radio emission comes from two giant clouds ejected from the radio galaxy's center.

Radio telescope A telescope that detects radio waves from objects in space.

Red giant An old star whose outer layers have billowed out and cooled down.

Red shift A shift in the spectral lines of most galaxies toward the red end of the spectrum. This is caused by the expansion of the Universe, which carries distant galaxies away from us.

Reflecting telescope A telescope that gathers light with a concave mirror.

Refracting telescope A telescope that collects light with a combination of lenses.

Relativity A theory devised by Albert Einstein that describes how objects move at very high speeds or in a powerful gravitational field.

Retrograde The apparent backward motion of a planet in the sky, which takes place because the Earth overtakes it in the Earth's own orbit.

Revolve To move in an orbit around a center of gravity. The Moon revolves around the Earth.

Rotate To spin on an axis. The Earth rotates on its axis once every 24 hours.

Satellite A small object that orbits a larger one; now mostly used for artificial satellites orbiting Earth.

SETI An acronym for the Search for Extra-Terrestrial Intelligence. This usually applies to searches for radio signals coming from other civilizations in the Universe.

Seyfert galaxy A spiral galaxy with a quasar-like core.

Solar flare An enormous explosion just above the surface of the Sun, caused by two magnetic loops touching.

Solar System Our Sun and its family of planets, moons, and assorted cosmic debris.

Solar wind The stream of energetic, charged particles blowing away from the Sun.

Solstice The two days in the year when the Earth's tilted axis causes the Sun to reach its maximum northern and southern positions above our planet. On June 21 the Sun reaches its northern extreme: it is the Summer Solstice for people living in the Northern Hemisphere, who then have their longest day and shortest night, and it is the Winter Solstice for people in the Southern Hemisphere. On December 22 the Sun is at its southern extreme in the sky, and the situation is reversed.

Spectroscope An instrument that is used to break up pure white light into its separate wavelengths.

Spectrum A band of radiation. For example, the rainbow colors that make up so-called white light, where each color corresponds to a different wavelength of radiation in the spectrum. Absorption lines in the spectrum of a star tell us about its chemical makeup.

Spiral galaxy A galaxy with spiral arms springing from a smooth central hub. Spiral galaxies have a mix of old and young stars, plus gas to form new generations of stars in the future.

Star A hot, massive, and luminous body that generates energy through nuclear fusion reactions.

Starburst galaxy A galaxy whose central regions have just undergone a sudden huge outburst of star formation, which is probably the result of an interaction with another galaxy.

Star cluster A cluster of stars. "Open clusters" are loose groups of a few hundred young stars; "globular clusters" are dense balls of almost a million old stars.

Sunspot A darker, cooler region on the Sun, caused by powerful magnetic fields stopping the normal circulation of gases. Sunspots build to maximum numbers and die away again in an 11-year solar cycle.

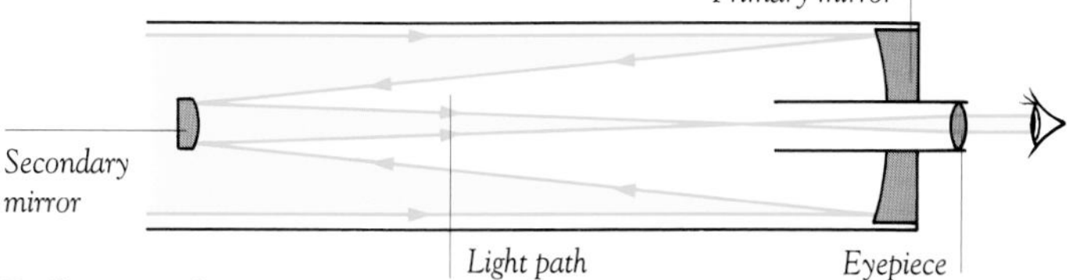

Reflecting telescope
This type of reflecting telescope is called a Cassegrain telescope, and it is the design on which the largest telescopes in the world are based. Starlight hits a concave mirror and bounces off a secondary convex mirror, before being brought to focus at the eyepiece.

Supercluster A cluster of clusters of galaxies.

Supernova A huge star that explodes at the end of its life.

Time zone A region of the Earth in which the official time is the same everywhere.

Ultraviolet Radiation with a wavelength that is shorter than ordinary light.

Umbra The dark, innermost zone of a sunspot. Also, the darkest part of a shadow caused by an eclipse.

Universe Everything that exists—stars, planets, galaxies, and dark matter.

Variable star A star that changes in brightness. Many variable stars regularly swell and shrink.

Velocity The speed of an object in one direction.

Wane A decrease in the phase of a moon or planet.

Wavelength The distance between the wave crests on any type of radiation (light, radio waves, etc.).

Wax An increase in the phase of a moon or planet.

White dwarf The collapsed core of a normal star like the Sun; all that is left after a star puffs off its outer layers as a planetary nebula.

X-rays Very short wavelength radiation that is produced by very hot gases.

Zenith The point in the sky directly overhead.

Zodiac The 12 star patterns against which the Sun, the Moon, and the planets appear to move.

Index

A

absorption line 96, 152
acceleration 52, 152
accretion disk 126–127, 152
Adams, John Couch 58
Albireo (star system) 121
Alcor (star system) 121
Aldebaran (star) 107
Aldrin, Buzz 54
Algedi (star system) 121
Algol (star system) 120
alien 146–147
Almach (star system) 120
Alpha Centauri (star) 116, 117, 121
Altair (star) 115
altitude 18, 28, 152
Andromeda
  (constellation) 120
  galaxy 136, 138
Antares (star) 107, 114
Apollo space missions
  Apollo 11 38, 54
  Apollo 12 37, 39
Aquila (constellation) 115
Armstrong, Neil 39, 54
asteroid 59, 152
asteroid belt 59, 86
astrolabe (stardial) 24–25, 152
astrology 26
astronauts
  on *Apollo* missions 54
  on *Mir* space station 34, 35
astronomers
  Beer, Wilhelm 48
  Bell, Jocelyn 127
  Bessel, Friedrich 116
  Drake, Frank 146
  Flamsteed, John 100
  Halley, Edmond 84
  Herschel, William 77, 133
  Hewish, Tony 127
  Hipparchus 114
  Hubble, Edwin 142
  Huygens, Christiaan 75
  Ihle, Abraham 135
  Leavitt, Henrietta 138
  Lockyer, Sir Norman 91
  Lower, Sir William 48
  Mädler, Johann Heinrich 48
  Maunder, E. W. 100
  Mayer, Tobias 48
  Messier, Charles 135
  Oort, Jan 133
  Ptolemy 58
  Tombaugh, Clyde 58, 80
  see also Galilei
astronomy 6–7
atmosphere 152
  of the Earth 16, 29
Atmosphere Probe 82, 83
atoms 152
Auriga (constellation) 115
auroras 17, 91, 152
  aurora australis 17
axis 42, 152

B

barycenter 40, 152
Beer, Wilhelm 48
Bell, Jocelyn 127
Bessel, Friedrich 116
Beta Centauri (star) 117
Beta Pictoris (star system) 123
Betelgeuse (star) 107
Big Bang theory 131, 142, 152
  Big Crunch and 144–145, 152
  inflation and 153
Big Dipper 24
binaries (star pairs) 120–121, 152
binoculars 11, 13
black holes 107, 126–127, 152
Braun, Wernher von 32
brown dwarf 152

C

cameras
  heat-sensitive 123
  at night 11
Capella (star) 115
Carina (constellation) 122
Cassini division 75
catadioptric telescope (cat) 12, 152
celestial sphere 152
Centaurus A (galaxy) 136
centrifugal force 72, 152
Cepheid stars 138, 152
Cernan, Eugene 39
Challenger space shuttle 17
Charon (Pluto's moon) 59, 80–81
chromosphere 99, 152
clock stars 24
clusters of galaxies 138, 152
Coal Sack 106
Collins, Mike 54
color
  of nebulae 123
  of sky 28
  of stars 118
Columbus, Christopher 38, 44
Coma Berenices (constellation) 149
comets 84–85, 152
  early beliefs on 59
  meteor showers and 86
computer imaging 137
Comte, Auguste 96
constellations 106, 110–111, 152
  Andromeda 120
  Aquila 115
  Auriga 115
  Carina 122
  Coma Berenices 149
  Crux 114, 117
  Cygnus 122
  Leo 110
  Orion 16, 31, 114
  Sagittarius 123
  Scorpius 16, 114
  Scutum 151
  Virgo 114
  zodiac 26–27
continental drift 17
Copernicus, Nicolaus 58
core (of a planet) 41, 152
corona 90, 152
cosmic ray 152
Cosmos 152; *see also* Universe
Crab Nebula 125
craters 49, 50, 152
crust (of a planet) 41, 152
Crux *see* Southern Cross
Cygnus (constellation) 122

D

dark adaptation 10
dark matter 131, 144–145, 152
Daylight Saving Time 22
degree 18–19, 152
Delta Aquarids (meteor shower) 86
Deneb (star) 90, 106
density 127, 152
Doppler, Christian 142
Doppler effect 142, 152
Drake, Frank 146
dust (cosmic) 118, 152

E

Earth 14–35
  distance from Moon of 40
  facts and figures 61
  inside 17, 41
  life on 16–17
  magnetic field of 17, 18
  Moon and 40–41
  orbit of (around Sun) 16, 26
  roundness of 19
  seasons on 26–27
  spin of 16, 20, 22
  time differences on 21
earthshine 46, 153
eclipse 120, 153
  of Moon 38, 44–45
  of stars 120, 121, 153
  of Sun 90–91, 102–103
eclipsing binary 120–121, 153
Einstein, Albert 64
electrons 96, 153
ellipse 62, 153
emission line 91, 153
Endeavour space shuttle 35
energy of Sun 92–93, 98–99
Epsilon Lyrae (star system) 121
equinox 153
equipment for home laboratory 8–9
ERS-1 satellite 35
escape velocity 32, 153
Eta Aquarids (meteor shower) 86
Eta Carinae Nebula 122
EURECA satellite 35
eyepiece 12, 153
extra-terrestrial intelligence 146–147

F

filaments and voids 131, 153
Flamsteed, John 100
flyby 153

focal length 12, 153
focus 12, 153
fossil fuels 92
fossils 16, 17
Fraunhofer, Joseph von 96
Fraunhofer lines 96–97, 152
frequency 46, 153
friction 87, 153

G

Gaia theory 17
galaxies 129–131, 153
  clusters of 138–139, 152
  elliptical 137, 153
  irregular 137, 153
  Large Magellanic Cloud 125, 131, 136
  Milky Way 132–135
  radio 131, 140, 155
  Seyfert 155
  Small Magellanic Cloud 136
  spiral 137, 155
  spotting 136
  starburst 131, 155
  types of 137
Galilei, Galileo
  on Copernican theory proof 58
  on Milky Way composition 133
  on Moon maps 48
  on Saturn's rings 74
  on Venus's phases 67
Galileo space probe 59, 82–83
gas giants 58–59, 153
Gaspra (asteroid) 59
Geminids (meteor shower) 86
geology
  of Mars 69
  of Moon 39, 54, 55
Giotto space probe 85
Global Positioning System (GPS) 22
globular clusters 135
Goddard, Robert Hutchings 32
gold 125
GPS 22
gravity 16, 52–53, 153
  acceleration due to 152
  forces between Earth and Moon 52
  on Moon 52, 54
  theory of 38
Great Dark Spot (Neptune storm) 78
Great Red Spot (Jupiter storm) 72, 73
greenhouse effect 153
  experiment 67
Greenwich mean time 21

H

Halley, Edmond 84
Halley's Comet 84
halos (lunar, solar) 28
Helix Nebula 124
Herschel, William 77, 133
Hewish, Tony 127
Hipparchus 114
Hubble, Edwin 142
Hubble Space Telescope 29
Huygens, Christiaan 75
Huygens space probe 75

I

IC 2391 (star cluster) 151
IC 2602 (star cluster) 150
Ihle, Abraham 135
inflation 144, 153
infrared 123, 153
International Date Line 21
International System of measurements (SI) 8
Io (Jupiter's moon) 59
ionosphere 29
IRAS satellite 123, 131

J

Jewel Box (star cluster) 150
Jupiter 72–73
  facts and figures 61
  Galileo space probe to 59, 82–83
  orbit time of 62
  size of 60
  spin speed of 72

K

Kapteyn, A. C. 133
Karl Zeiss company 109
Keck Telescope 29
Kepler, Johannes 63
Kitt Peak Observatory 31
Krikalev, S. 35
Kuiper Belt 84, 153

L

La Palma Observatory 31
Lagoon Nebula 123
Landsat satellite 15
Large Magellanic Cloud 125, 131, 136
Las Campanas Observatory 130
latitude 18–19, 153
  measuring 18, 19
  see also longitude
Leavitt, Henrietta 138
Leo (constellation) 110
Leonids (meteor shower) 86
light pollution 30–31, 153
light spectrum 96
light-years 107, 117, 153
Lockyer, Sir Norman 91
Long, Roger 109
longitude 18–19, 153
  measuring 20, 21
  see also latitude
Lower, Sir William 48
luminous glowing 154
Luna 2 space probe 54
Luna 3 space probe 39
Lunar Orbiter space probes 48
Lyrids (meteor shower) 86

M

M5 (globular cluster) 135
M13 (globular cluster) 135
M22 (globular cluster) 135
M32 (galaxy) 137
M33 (galaxy) 136, 138
M34 (star cluster) 149
M41 (star cluster) 150
M47 (star cluster) 150
M82 (galaxy) 131
M83 (galaxy) 137
McCandless, Bruce 17
McMath Solar Telescope 90
Mädler, Johann Heinrich 48
Magellan spacecraft 66
magnetic field 154
  on Earth 17, 18
  on Sun 90, 100
magnetosphere 154 (*see* magnetic field)
magnitude 114, 154
mantle (of planet) 41, 154
mare, maria 48, 154
Mariner 10 spacecraft 64
Mars 68–71
  "canals" on 68, 69
  dust storms on 71
  facts and figures 61
  moons of 68
  orbit time of 62
  size of 60
  Viking space probes 68, 70
Mauna Kea (volcano) 68
Maunder, E. W. 100
Mayer, Tobias 48
measurements
  rounding off 10
  SI (International System) 8
Mercury 64–65
  facts and figures 61
  orbit time of 62, 64–65
  size of 60
mesosphere 29
Messier, Charles 135
meteor (shooting star) 11, 86–87, 154
meteor showers 11, 86–87, 154
meteorites 86, 154
  lunar craters and 49, 50
Meudon Observatory 31
microwave background 154
Milky Way 132–133
  structure of 134–135
Mir space station 34, 35
Miranda (Uranus's moon) 76
Mizar (star system) 121
molecule 154
Moon 36–55
  Apollo space missions to 37, 38, 39, 54
  craters on 49, 50
  distance from Earth of 40
  Earth and 40–41
  earthshine on 46, 153
  eclipse of 38, 44–45
  geology of 39, 54, 55
  gravity on 52, 54
  halos of 28
  inside 39, 41
  mapping of 48–49
  maria of 49, 154
  mountains on 50–51

orbit of (around the Earth) 42–43
origins of 39, 51
phases of 42–43
spotting of 38, 46–47
tides and 52–53
moonbase 39, 55
moondial 47
moonrocks 39, 54, 55
moons 38, 154
of Jupiter 38, 72
of Mars 68
of Neptune 38, 78
of Pluto 59, 80–81
of Saturn 38, 74
of Uranus 76

N

navigation using stars 16, 18, 19
star theaters 108
Navstar satellite 22
nebulae 122–123, 154
planetary 124
Tarantula Nebula 131
Neptune 78–79
discovery of 58
facts and figures 61
orbit time of 62
size of 60
neutron stars 126–127, 154
Newton, Isaac 38, 52
NGC 253 (galaxy) 136
NGC 1365 (galaxy) 137
NGC 2362 (star cluster) 150
nocturnal (instrument) 154
North America Nebula 122
north (finding due north) 22
North Star *see* Polaris
nova (*see* white dwarfs) 154
nuclear fusion 98, 99, 154
nucleus 154

O

observatories 154
home 10–11
Kitt Peak (Arizona) 31
La Palma (Canary Islands) 31
Las Campanas (Chile) 130
light pollution and 30, 31
Meudon (Paris, France) 31
Royal Observatory (Greenwich, U.K.) 21
occultation 154
Olympus Mons (Martian volcano) 68
Omega Centauri (star) 135
Omega Nebula 123
Oort Cloud 84, 133, 153, 154
Oort, Jan 133
orbit 59, 154
Orion (constellation) 16, 31, 114
Orion Nebula 75, 105, 123
Orionids (meteor shower) 86
orrery 59
ozone layer 29

P

parallax 116, 154
parhelia 28
Pelican Nebula 122
penumbra 45, 154
Penzias, Arno 131
Perini's Planetarium 109
Perseids (meteor shower) 86
phase 38, 154
photography of sky 11
with heat-sensitive cameras 123
star trails and 24
photosphere 99, 154
Pioneer Venus Orbiter spacecraft 67
Planet X 80–81
planetariums 108, 109, 154
planets 58–59, 154
facts and figures 61
formation of 59, 119, 123
orbits of 62–63
sizes of 60–61
see also individual planets
planispheres 112–113
plate tectonics 17
Pleiades (star cluster) 106, 148
Pluto 80–81
facts and figures 61
moon of 59, 81
orbit of 62, 80
size of 60
Polaris (Pole Star) 154
finding 24
and latitude 18, 19
and timekeeping 24, 25
pollution and the atmosphere 29
Praesepe (star cluster) 149
prime number 146, 154
prism 96, 154
prominence 90, 154
proton 99, 154
Proxima Centauri (star) 90, 91, 107, 116, 117
Ptolemy 58
pulsars 126–127, 154

Q

Quadrantids (meteor shower) 86
quasars 131, 140–141, 154

R

radar 66, 154
radar mapping 66–67
radiant 99, 154
radiation damage 29
radio galaxies 131, 140, 155
radio signals from alien civilizations 146–147
radio telescopes 130, 155
rainbows 96
red giants 107, 118–119, 155
red shift 116, 155
reflecting telescope 12, 154
refracting telescope 12, 154
relativity theory 64, 155
retrograde motion 62–63, 155
revolve 155
rockets 32–33
building 33
see also space missions
Rosetta space probe 85
rotate 40, 155
Royal Observatory (Greenwich, U.K.) 21

S

Sagittarius
constellation 123
star clusters 151
satellites 32, 34–35, 155
ERS-1 35
EURECA 35
IRAS 123, 131
Landsat 15
Navstar 22
spotting 35
Saturn 74–75
facts and figures 61
orbit time of 62, 74
size of 60
Saturn V moon-launcher 32
Scooter (Neptune storm) 78
Scorpius
constellation 16, 114
star clusters 15
Scutum (constellation) 151
SETI (Search for Extra-Terrestrial Intelligence) 146–147, 155
Seven Sisters (Pleiades star cluster) 106, 148
Seyfert galaxy 155
shooting stars 11, 86–87
SI (International System of measurements) 8
silver 125
Sirius (star) 115
sky color 28
Small Magellanic Cloud 136
solar eclipse 90–91, 102–103
solar flares 90, 100, 155
Solar System 57–87, 155
planet facts and figures 61
planet sizes 60–61
see also individual planets; Moon; Sun
solar wind 91, 155
solstice 90, 155
Southern Cross (Crux) 114, 117
binary star systems in 121
finding latitude with 18, 19
space missions
Apollo series 37, 38, 54

Atmosphere Probe 82, 83
Challenger space shuttle 17
to comets 85
Endeavour space shuttle 35
escape velocity in 32, 153
flyby and 153
Galileo space probe 59, 82–83
Giotto space probe 85
Huygens space probe 75
interstellar spacecraft to 17
Jupiter 82–83
Luna space probes 39, 54
Lunar Orbiter space probes 48
Magellan spacecraft 66
manned lunar landings 39, 54
Mariner 10 spacecraft 64
to Mars 68, 70–71
to Neptune 78
past, present, and future 33
rocket pioneers and 32
Rosetta space probe 85
to Saturn 75
to Uranus 76
to Venus 66, 67
Viking landers 68, 70
Voyager space probes 57, 74, 76, 78
space probes
see space missions
space shuttle 33, 34
Challenger 17
Endeavour 35
space stations 34–35
spectroscope 97, 155
spectrum 96, 155
Spica (star) 107
spiral galaxy 137, 155
star charts 11
making a planisphere 112–113
star-dial 25
stars 104–127, 155
birth of 107, 119, 122–123
Cepheid 138, 152
clock 24
clusters 150, 151
constellations 26–27, 106, 110–111
daytime invisibility of 29
death of 119, 124–127
distance from Earth 116–117
double (binaries) 120–121
eclipsing binary 120, 121, 153
making a planisphere and 112–113
making a star theater 108–109
of northern skies 148–149
pulsars, black holes and 126–127
red giant and white dwarf 107, 118–119, 155
shooting 11, 86–87
of southern skies 150–151
supernovae (star explosions) 125, 155
temperatures of 118
twinkling of 114–115
variable 115, 155
starspots 95
Stonehenge, U.K. 90
stratosphere 29
Sun 88–103
brightness of 64, 93
corona of 90, 152
cycle of 100–101
death of 124
eclipse of 90–91, 102–103
energy of 92–93, 98–99
halos of 28
inside 91, 96, 98–99
projector 94–95
studying 90
see also solar flares; solar wind; sundial; sunlight; sunspots
sundial 22
finding due north with 22
making a 22–23
sunlight 96–97
sky color and 28
time to reach the Earth 98
sunsets 28
sunspots 89, 100–101, 155
cycle of 100–101
measuring 95
movement of 94
projector 94–95
superclusters 131, 155
supernovae 125, 155
pulsars and 126, 127

T

Taurids (meteor shower) 86
telescope 12–13, 155
catadioptric (cat) 12, 152
Hubble Space 29
IRAS 123
Keck 29
McMath Solar 90
making 12–13
radio 130
reflecting 12–13, 155
refracting 12–13, 155
Very Large Array radio 130
X-ray 95
Theory of Relativity 64
thermosphere 29
tides 52–53
time
of day 22–23
differences 21
lunar calendars and 42, 43
at night 24–25
star-dials and 25
sundials and 22–23
zone 21, 155
Titan (Saturn's moon) 59, 74, 75
Tombaugh, Clyde 58, 80
Trifid Nebula 105, 123
Triton (Neptune's moon) 59, 78, 79
troposphere 29
Tsiolkovsky, Konstantin 32
47 Tucanae (star cluster) 130, 135

U

ultraviolet 91, 155
umbra 45, 155
Universe 129–147, 155
aliens 146–147
Big Bang theory of 131, 142, 144–145, 152
expanding 142–143
galaxies in 130–137
uranium 125
Uranus 76–77
facts and figures 61
orbit time of 62, 76
size of 60
Ursids (meteor shower) 86

V

variable stars 115, 155
Vega (star system) 123
velocity 32, 155
Venus 66–67
facts and figures 61, 66
orbit time of 62
phases of 67
size of 60
Very Large Array radio telescope 130
Viking space probes 58, 68, 70
Virgo
constellation 114
galaxy cluster 139
voids and filaments 131, 153
volcanoes 17
formation of 68
on Mars 68
Volkov, V. 35
Voyager space probes 57, 74
Voyager 2 76, 78

W

wave 10, 155
wavelength 91, 155
wax 155
weather
on Jupiter 72, 73
on Mars 71
on Neptune 78
sunspot activity and 100
white dwarfs 107, 118, 155
Wild Duck (star cluster) 151
Wilson, Robert 131

X

X-rays 91, 95, 155

Y

Young, John W. 54

Z

Zeiss company 109
zenith 155
zodiac 26–27, 155
zodiacal light 87

Acknowledgments

HEATHER COUPER and NIGEL HENBEST would like to thank Robin Scagell, Beverley Miles, Frank Miles, Helen Sharman, and Brian Stockwell for their assistance.

DORLING KINDERSLEY would like to thank Jack Challoner and Fran Halpin for consultancy; Sarah Moule for picture research; Louise Abbott, Emily Hill, and Alison Woodhouse for editorial assistance; Karin Woodruff for the index; Heather Couper and David Pelham for permission to use their red shift and dialing the Universe concepts as the basis for experiments; and the YHA Adventure Shop, London, for providing the camp stove.

PHOTOGRAPHY by Andy Crawford, Tim Ridley, and Steve Gorton. Photographic assistance from Gary Ombler, Nick Goddall, and Sarah Ashun. Additional photography by Colin Keates, Dave King, and Clive Streeter.

PICTURE CREDITS
t top; *c* center; *b* below; *r* right; *l* left; *a* above
Ashford School, Kent: 35*bl*, 35*c*.
A.T.& T. Archives: 131*br*.
Jean-Loup Charmet: 48*tca*, 48*tcb*, 48*tr*.
Bruce Coleman Ltd. / Dr. E. Potts: 86*bl*.
E.T. Archive: 19*bl*, 106*bl*; / *Uffizi Gallery, Florence*: 72*tr*.
Mary Evans Picture Library: 63*tl*, 77*cr*, 84*bl*, 98*tr*, 109*cr*, 146*br*.
Galaxy Picture Library: 90*tl*, 91*tr*; / *NASA*: 59*tr*, 123*c*; / *R. Scagell*: 11*cr*, 11*br*, 35*cr*.
Robert Harding Picture Library / G. Hellier: 17*cl*.
Hencoup Enterprises: 58*tl*, 100*tr*, 127*tr*; / *NASA*: 59*bl*; / *C. Tombaugh*: 58*bl*.
Michael Holford: 22*bl*.
The Image Bank: 16*bl*; / *J. Rajs*: 46*tr*.
Images Colour Library: 38*bl*, 43*tr*.
Ann Ronan at Image Select: 44*tr*, 130*tl*.
Frank Lane Picture Agency / Hosking: 103*bl*.
Leiden Observatory / L. Zuyderduin: 133*tl*.
Mansell Collection: 75*cr*, 116*tr*.
NASA: 36–37, 37*tr*, 54*cr*, 59*tl*, 67*tl*; / *NASA Jet Propulsion Laboratory*: 74*tr*.
Novosti, London: 39*tl*.
Oxford Scientific Films / D. Thompson: 52*tr*, 52*c*.
Popperfoto: 64*br*.
Science Graphics, Inc.: 62*tr*.
Trustees of the Science Museum, London: 24*br*.
Science Photo Library: 58*tr*, 90*tr*, 142*bl*; / *J. Baum*: 107*b*; / *Dr. J. Burgess*: 38*br*; / *Dr. R. Clark & M. Goff*: 123*tr*; / *Dr. F. Espenak*: 38*tc*; / *G. Fowler*: 128–129; / *G. Garradd*: 17*cr*, 24*tr*; / *G.E. Astro Space*: 21*br*; / *Dr. L. Golub*: 95*tr*; / *K. Gordon*: 124*tr*; / *Hale Observatories*: 95*tl*; / *D. Hardy*: 56–57; / *Dr. J. Lorre*: 131*bl*, 137*tr*; / *J. Mead*: 93*tr*; / *D. Milon*: 71*tl*; / *NASA*: 14–15, 16*tl*, 17*br*, 32*tr*, 32*c*, 38*tr*, 39*br*, 40*tr*, 66*cl*, 68*tr*, 76*tr*, 88–89; / *NOAO*: 104–105, 130*br*, 137*br*; / *Novosti*: 32*tl*, 34*tr*; / *P. Parviainen*: 86*tl*; / *G. Post*: 91*b*; / *R. Ressmeyer, Starlight*: 31*tr*, 130*tr*, 130*bl*; / *Royal Greenwich Observatory*: 125*tr*, 137*bl*; / *Royal Observatory, Edinburgh*: 35*cl*, 106*t*, 125*br*, 131*t*, 137*cra*, 139*br*; / *Royal Observatory, Edinburgh / AATB*: 105*tl*, 137*crb*; / *Rev. R. Royer*: 84*tr*, 106*br*, 107*tl*, 133*br*; / *J. Sanford*: 28*tr*, 87*tr*; / *R. Scagell*: 117*tr*; / *U.S. Naval Observatory*: 233*br*; / *K. Wood*: 90*b*.
Space Imagery Center, Lunar and Planetary Laboratory, University of Arizona, Tucson: 49*bc*, 49*br*.
TASS: 35*br*.

Every effort has been made to trace the copyright holders, and we apologize in advance for any unintentional omissions. We will be pleased to insert the appropriate acknowledgment in any subsequent edition of this publication.

ILLUSTRATIONS
Kuo Kang Chen: 18, 21, 22, 24–25, 26, 28, 30, 31, 40, 42, 44, 45, 46, 50, 52, 54, 64, 66, 68, 72, 74, 76, 78, 80, 83, 92, 94, 96–97, 98, 100, 102, 103, 108, 110, 114–115, 116, 118–119, 120, 121, 122–123, 124, 135, 136, 148–149, 150–151, 152–153, 154–155.
Mark Franklin: 12, 33, 49, 95, 126, 140, 142–143.
Additional illustrations by Luciano Corbella, Martyn Foote, and Chris Lyon.

MODEL MAKING
Peter Griffiths: 13, 21, 23, 25, 27, 31, 36, 43, 45, 47, 68, 72, 74, 76, 80–81, 82–83, 85, 101, 102, 109, 110–111, 120, 127, 134, 139, 141.
Mark Franklin: 49, 53, 60–61, 63, 67, 116, 143, 144, 146.

MODELS
Sarah Ashun, James Bitmead, Rebecca Bunting, Tom Bunting, Rhiannon Bushnell, Jonathan Chen, Zane Cunningham, Charlotte Denton-Cowell, Rosie Denton-Cowell, Lia Foa, Maya Foa, Steve Gorton, Georgina Grant, Peter Griffiths, Stephanie Jackson, Hannah Janulewicz, Jan Janulewicz, Darius Jones, Wesley Lincoln, Elaine Monaghan, Carly Nichols, Marianna Papachrysanthou, Anita Parsons, Elizabeth Parsons, Priya Patel, Alastair Raitt, Duncan Raitt, Samantha Schneider, Matthew Smedley, Kerry-Anne Smith, Kathy-Marie Smith, Mimy Tang, Joanna Temple, Andrew Thomas, Calvin Thomas, Nicholas Turpin, Ailsa Williams, Andy Williams, Scallywags Agency.